IMPROVING
SCIENCE
EDUCATION

THE NATIONAL SOCIETY FOR THE STUDY OF EDUCATION
The Series on Contemporary Educational Issues

IMPROVING SCIENCE EDUCATION

Edited by
Barry J. Fraser and Herbert J. Walberg

Editor for the Society
Kenneth J. Rehage

19 NSSE 95

Published in cooperation with
THE INTERNATIONAL ACADEMY OF EDUCATION

Distributed by THE UNIVERSITY OF CHICAGO PRESS • CHICAGO, ILLINOIS

The National Society for the Study of Education

The National Society for the Study of Education was founded in 1901 as successor to the National Herbart Society. The purpose of the Society is to encourage serious study of educational problems and to make the results of such study available through its publications.

These publications include the annual two-volume Yearbook, the first of which was published in 1902. The two volumes of the 94th Yearbook, published in 1995, are entitled *Creating New Educational Communities* and *Changing Populations/Changing Schools*.

In 1972 the Society established a new series of publications under the series title "Contemporary Educational Issues." The present volume, *Improving Science Education*, is one of two published in that series in 1995. The other volume is entitled *Ferment in Education: A Look Abroad*.

All members of the Society receive its two-volume Yearbook. Members who take the Comprehensive Membership also receive the two current volumes in the series on Contemporary Educational Issues.

Membership in the Society is open to any individual who desires to receive its publications. Inquiries regarding membership and current dues may be addressed to the Secretary-Treasurer, NSSE, 5835 Kimbark Avenue, Chicago, IL 60637.

Library of Congress Catalog Number: 95-67502
ISBN: 0-226-260984

Published 1995 by

THE NATIONAL SOCIETY FOR THE STUDY OF EDUCATION
5835 Kimbark Avenue, Chicago, Illinois 60637

First Printing

Printed in the United States of America

v

Foreword

The International Academy of Education is a not-for-profit scientific association that promotes educational research, its dissemination, and the implementation of implications from its findings. The Academy is dedicated to strengthening the contributions of research, solving critical educational problems throughout the world, and providing better communication between researchers and practitioners. One of the main activities of the Academy is the organization of international task forces to produce reviews that synthesize research on effective educational practices and policies. The resulting reports are intended to represent the best thinking and critical review that can be brought together from several countries. The present publication is the outcome of one of these task forces.

This book encompasses an area of education—namely, science education—which is recognized widely as being of great importance internationally both for the economic well-being of nations and because of the need for a scientifically literate citizenry. Essentially the book's aim is to contribute to the improvement of science teaching and learning at all levels of education.

The book is organized so that each chapter is written by one or more international authorities and focuses on a different topic of central importance in science education (e.g., science curricula, students' conceptions, instructional strategies, assessment and evaluation, classroom environments, teacher change, use of computers, and greater equity). Also, each chapter is research-based in that it draws upon research findings from studies undertaken around the world. Finally, each chapter is written in a relatively nontechnical way and provides decision makers with practical (yet research-based) suggestions for improving science education.

This publication attempts to come to grips with important issues and problems in science education. It is one of several state-of-the-art reports planned when the Academy was established in 1986. The Academy, which consists of leading scholars in academic disciplines relevant to education, decided from the outset that such reports should address policy issues common to many countries.

I would like to thank the distinguished group of scholars whom the Academy invited to form a task force and to meet at Curtin University in Perth, Australia to plan the book: Herbert Walberg of the University of Illinois at Chicago (Chair), John Keeves (Flinders University of South Australia), George Marx (Roland Eotvos University, Hungary), K. C. Cheung (now at the University of Macau), Wayne Welch (University of Minnesota, USA), Sheila Haggis (UNESCO, France) and Barry Fraser (Curtin University of Technology, Australia). Also, the Academy is grateful to the chapter consultants from around the world for providing valuable feedback which enhanced the quality of the book. Last, but not least, I want to express gratitude on behalf of the Academy to the book's editors and chapter authors for a significant, illuminating, and scholarly work on the internationally important topic of science education.

Gilbert de Landsheere
Chairman, International Academy of Education
Liège, Belgium, October 1994

Acknowledgments

As is apparent in the "Foreword," this volume is the result of an initiative undertaken by the International Academy of Education. In 1993, when work on the volume was well along, the editors inquired about the possibility of its being published as one of the volumes in the NSSE Series on Contemporary Educational Issues. The committee responsible for planning that series responded warmly to this suggestion and it was agreed that publication would be scheduled for 1995.

As the reader will note, authors of chapters in this volume and the consultants who reviewed the manuscripts submitted come from all over the world. We are grateful to the authors and consultants for their contributions.

We are especially indebted to the editors, Professors Barry J. Fraser and Herbert J. Walberg, for their painstaking work in the initial editing of the manuscripts as well as their part in planning the volume. Both have assisted greatly at every stage of seeing the book through to publication.

Kenneth J. Rehage
Editor for the Society

Table of Contents

Chapter 1

INTRODUCTION AND OVERVIEW

Herbert J. Walberg and Barry J. Fraser

In accordance with the title of this volume, *Improving Science Education*, the purpose of this book is to specify how elementary and secondary science education can be improved. We hope that this purpose has been accomplished by the reviews of research literature by distinguished specialists in science education, and by the practical implications of the research that have been drawn out explicitly.

Clearly, no country has a monopoly on effective policies, practices, and research in this field. The scholars and practitioners who prepared chapters for this book, and the consultants who assisted them, have been drawn from many countries outside the West, including Africa and Asia. Further, we hope that people in various parts of the world will find helpful the insights and research of their professional peers in other countries. Perhaps science education can become more scientific and universal in the way that physics and chemistry are.

In this introductory chapter we discuss the current importance given to science education worldwide and provide an overview of the chapters that follow.

CURRENT IMPORTANCE OF SCIENCE EDUCATION

According to Paul DeHart Hurd, pressures for the reform of science education have been underway since 1970, when it became clear that projects to improve course content in the 1960s were at best disappointing (Hurd, 1993a). During the 1980s and 1990s, there appeared approximately

400 reports which advocated reform in general, with many specifically highlighting science education reform (Hurd, 1994). Some of these pressures for reform came from the change from an industrial age to a knowledge-intensive era, changes in science itself, and the implications of research in cognitive science for curriculum and instruction. According to UNESCO, there are 141 countries currently revising school science curricula, and the only common goal in all countries is "learning to learn" (Hurd, 1994). Despite these reform efforts, however, Hurd (1993b, p. 1009) claims that "the movement has been propelled by a massive onslaught of rhetoric, an overload of undefined platitudes, political exaltations, unenlightened statistics, finger pointing, and educational policies based on intuition at best."

The well-known *A Nation at Risk: The Imperative for Educational Reform* report (National Commission on Excellence in Education, 1983) is often considered to mark the first of several main waves of reform. Early reforms emphasized a back-to-basics thrust as politicians and business people called for an emphasis on the three Rs and more stringent graduation requirements. Not surprisingly, the top-down nature of the first wave of reforms attracted skepticism (Cuban, 1984). Later, reform policy moved from an emphasis on the basics to more radical restructuring of the curriculum to emphasize active learning, problem-solving, application of technology, interpersonal skills, and relating science to the everyday life of the student (Odden and Marsh, 1987). In particular, higher-order thinking skills (Tobin, Kahle and Fraser, 1990) grew in importance.

Science education has occupied a central place within these pushes for reform. For example, in *Educating Americans for the 21st Century* (National Science Board Commission on Precollege Education in Mathematics, Science, and Technology, 1983), there was a call for American students to be world leaders in science and mathematics achievement by the year 1991. In Australia, the Federal Government established a Key Center for School Science and Mathematics in 1988 to stimulate the improvement of science and mathematics teaching (Fraser, 1994). Also, this was a period when many dozens of reports related to reform in science appeared, including *Fulfilling the Promise: Biology Education in the Nation's Schools* (National Research Council, 1990), and *Changing America: The New Face of Science and Engineering* (Task Force on Women, Minorities and the Handicapped in Science and Technology,

1989). During this period, reports of the National Center for Improving Science Education (1989) recommended desired curriculum changes for elementary, middle, and high school, including focusing on fewer topics at greater depth. Moreover, the emphasis on reform of science education at the national level is paralleled at the state level in numerous states including Florida (Task Force for Improving Mathematics, Science, and Computer Education, 1989).

The National Science Teachers Association's (1989) report, *Essential Changes in Secondary Science: Scope, Sequence, and Coordination,* provides recommendations about how to coordinate the scientific ideas in the curriculum at different grade levels. Topic sequences for grades 6-8, 9-10, and 11-12 are provided as templates to guide the design of new curricula (National Science Teachers Association, 1992). Essentially, the content of science is organized according to the tenets of scope, sequence, and coordination. However, these guidelines are flexible and capable of responding to varied needs and situations.

Perhaps the science education reform effort which has received the most attention worldwide in recent times in *Project 2061: Science for all Americans* (AAAS, 1989). This report emphasized the importance of cultivating a scientifically literate society and urged substantial and systematic changes in traditional science curricula (Rutherford and Ahlgren, 1990). Although not intended as a curriculum guide, the goals set out in Project 2061 are being used in this way in several American states (Project 2061, 1992) and have stimulated much thinking and debate about science education reform. The recent *Benchmarks for Science Literacy* (AAAS, 1993) specifies how students should progress toward scientific literacy and recommends what they should know and be able to do at certain grade levels.

Another major current project, the *National Science Education Standards* (National Science Council, 1994), offers a coherent vision of what it means to be scientifically literate. The standards describe what all students must understand and be able to do as a result of their science education. They also provide criteria for judgments regarding teaching and programs. The standards are organized into five categories: science system standards which specify the support systems and resources needed to provide all students with the opportunity to learn science; science program standards which specify the nature, design, and consistency of school science programs; science teaching and professional development standards which specify the criteria for the exemplary practice of science teaching; science assessment standards which specify criteria for assessing and analyzing student attainments in science; and science content standards which specify expectations for the development of proficiency in conducting inquiry, the ability to apply and communicate scientific knowledge, scientific understandings, understanding of the relationship between science and technology, and the influence of science on societal issues.

For the first time in the history of the field of science education, a prolific number of studies has been synthesized in a comprehensive manner in handbooks. *The Handbook of Research on Science Teaching and Learning* (Gabel, 1994), which was sponsored by the National Science Teachers Association in the USA, encompasses 19 chapters divided into four sections. The two-volume *International Handbook of Science Education* (Tobin and Fraser, in preparation) encompasses over 70 chapters and is divided into 10 sections.

OVERVIEW

In "Science in a Changing World" (chapter 2), John Keeves and Glen Aikenhead begin by briefly characterizing the history of school science education and analyzing the distinctive characteristics of school science in several countries. They describe the curriculum reforms of the mid-twentieth century, including the modernization of chemistry and physics, the emergence of biology as a coherent field, renewed emphasis on laboratory work, the introduction of elementary school science, and reduced emphasis on applications. Drawing on cross-national research, they report participation rates, conditions for learning science, and emphasis on different content and process objectives. Keeves and Aikenhead then report an extensive analysis of curriculum changes of the 1980s around the world, such as universal secondary education, life-long education, learning to learn, and science-related social issues. They conclude with a discussion of the long-range goals for science education being embodied in current curriculum efforts. In order to improve science education, Keeves and Aikenhead recommend that:

- all students should formally study some science during all grades of schooling;

- science courses should emphasize content from the four fields of science (biology, chemistry, earth science, and physics), the skills of investigation and inquiry, and relevance to everyday life;
- relationships between science, technology, and society should be emphasized;
- the learning of science should be seen as a lifelong process, not limited to the years of schooling.

In "Students' Conceptions and Constructivist Teaching Approaches" (chapter 3), Reinders Duit and David Treagust present a new view in the psychology of teaching of particular concern in science teaching. Although some students are able to answer questions correctly about science in the context of their academic studies, they are unable to apply their understanding to real-world problems and even might not believe the academic explanation of phenomena. Influenced by philosophers from Plato to Piaget, constructivist investigators believe that knowledge must be created rather than imbibed by the student. To attain this knowledge, students must be confronted with the failure of their false conceptions to explain common phenomena. Duit and Treagust describe research that illustrates this view, and recommend to policymakers and administrators that:

- science teaching should start with the student's conceptions rather than the teacher's or scientist's view;
- curricula should consider students' conceptions and present science in a social context;
- assessment procedures should emphasize understanding rather than recall;
- new media, such as computer software, should be used to provide students with opportunities to construct their own knowledge.

In "Instructional Strategies" (chapter 4), Avi Hofstein and Herbert Walberg describe the findings and implications of research on effective instructional strategies in school science. They point out that a traditional goal in school science has been the teaching of inquiry skills in an attempt to impart the excitement of scientific discovery and a spirit of independent investigation. Nonetheless, it has been difficult to attain this goal and to demonstrate beneficial effects of laboratory work. Although computer simulations might not have a much better record of success, they take less time than laboratory work and are as well liked by students. Distance education is another relatively new and effective technique that increases opportunities and saves time and costs. Some other instructional strategies discussed by Hofstein and Walberg are direct teaching, teacher demonstrations, and field trips. The authors recommend that:

- science teaching should involve a variety of instructional techniques, including well-designed teacher's demonstrations, laboratory work, computer simulations, and field trips;
- student outcomes should be assessed thoroughly both in the classroom and in the laboratory;
- distance education should be used as an effective method in both high-income and low-income countries;
- teachers should be trained in using a 'wait-time' of appropriate duration.

In "Student Assessment and Curriculum Evaluation" (chapter 5), Wayne Welch begins with a brief recounting of the history of science reform together with its assessment and evaluation implications. Reflecting fears of student alienation from science, one major trend is the assessment of attitudes toward science. Influenced by Piagetians and other constructivists, psychologists and evaluators also have become increasingly interested in the measurement of thinking and higher cognitive skills. Because governments increasingly seek value for money, evaluators have investigated the effects of many curricula and programs. Many new means of assessment have evolved beginning with the reforms of the 1960s (e.g., clinical interviews). In order to improve science education, Welch recommends that:

- emphasis on the evaluation of student learning should be supplemented by greater attention to formative and summative curriculum evaluation;
- the focus on student achievement should be extended to encompass attitudes, inquiry skills, problem-solving, and understanding of the nature of science;
- paper-and-pencil evaluation instruments should be complemented by the use of alternative and authentic evaluation techniques such as in-depth interviews, portfolio assessments, and practical performance tests;

- the usefulness of formative curriculum evaluation should be improved by seeking highly specific information that can guide improvements and by using separate student and curriculum evaluation instruments.

In "Classroom Learning Environments" (chapter 6), Barry Fraser and Theo Wubbels describe the nature and measurement of learning environments which can be defined as the perceived quality of the social context of learning. The quality of the classroom environment is important to teachers and policymakers wanting to know the comparative merits of new curricula and various teaching methods such as laboratory work, direct instruction, simulation, and field trips. Fraser and Wubbels describe what it is about students' educational experience that leads to achievement and positive attitudes. A description is given of how teachers can use assessments of classroom environment to guide improvements in their classrooms. Implications of learning environment research for improving science education are that:

- student perceptions of learning environment should be used to provide information about subtle but important aspects of classroom life;
- science teachers should strive to create 'productive' learning environments that emphasize more organization, cohesiveness, and goal direction, and that have a greater similarity between the actual environment and that preferred by the class;
- the evaluation of innovations and new curricula should include classroom environment assessments;
- teachers should use assessments of their students' perceptions of actual and preferred classroom environment to monitor and guide attempts to improve classrooms.

In "Teacher Change and the Assessment of Teacher Performance" (chapter 7), Kenneth Tobin distinguishes the 'objectivist' assumption that teachers can assimilate objective subject matter and pedagogical knowledge from the contrary view that teachers must create or construct their knowledge from experience of the subject matter and teaching it. Tobin shows that early research revealed that many useful effects could be obtained by objectivist methods of feedback, microteaching, and behavioral analysis. In addition, it appears that coaching and metaphorical analysis in line with constructivist views also can have useful effects in encouraging science teachers to reflect critically about their work and to change their practices. Tobin recommends that:

- teachers should observe one another's classes and discuss what did and did not work;
- teachers should conceptualize new teaching roles in terms of metaphors, analyze curricular events in terms of beliefs about power relationships, and identify constraints that prevent teachers from implementing the curriculum as they would prefer;
- professional practice schools should be developed as collaborative partnerships between schools, universities, and the community;
- portfolios and simulations should be used in the assessment of teacher performance.

In "Use of Computers" (chapter 8), Tjeerd Plomp and Joke Voogt describe philosophies of science education expounded during the past three decades beginning with inquiry, proceeding through to a societal emphasis, and then to current constructivist emphasis on conceptual change. They argue that computer-managed and computer-assisted instruction, along with simulation and other technologies, can help to accomplish the goals of these several philosophies. From the world's most extensive survey, they describe how computers and other technologies currently are being used in 19 countries. Although much computer use is for didactic purposes, many teachers around the world are employing simulation, graphics, statistics, and spreadsheet software for teaching science. Although it appears that many computer-using science teachers are enthusiastic about the computer's potential for further improvement, problems are yet to be overcome in integrating computer-based material into the regular curriculum. The authors recommend that:

- the introduction of computers should be accompanied by a support structure which incorporates competence development, the training of consultants, the provision of network facilities, and the building of organizational capacity;
- teachers inexperienced in computer use in the classroom should receive in-school training in which they can observe good examples of educational computer use and have hands-on experience;
- courseware should support the existing goals of the curriculum and include advice on its appropriate use;

- attention should be given to formative evaluation of courseware in classroom settings as a basis for making needed improvements.

In "Gender Equity" (chapter 9), Lesley Parker, Léonie Rennie, and Jan Harding describe the differences between boys and girls in science attitudes, learning, and participation. In general, boys do better in all three respects, partly because of instructional inequality. For example, an analysis of 81 studies of teacher-pupil interaction showed that girls receive less of the teacher's attention in classes under a wide range of conditions. Of some 600 programs for extending better opportunities to girls, many rely on segregated classes, special curricula, teacher education, role modeling, and extra-curricular and co-curricular activities. The authors discuss the conditions for success of such programs, and their recommendations for improving gender equity are that:

- the school science curriculum should be compulsory, broad-based, and include consideration of gender stereotypes and career education aimed at breaking down these stereotypes;
- resources in science education should avoid gender bias in language and choice of examples, and include case studies of successful women scientists;
- science teachers, administrators, teacher educators, and teacher trainees should undertake educational programs that make them aware of the problems of gender stereotyping and give them the skills to counter it;
- gender bias should be avoided in the content, context, and mode of assessment in science.

In "Cross-National Comparisons of Outcomes in Science Education" (chapter 10), John Keeves reports the findings of an extensive international survey of science participation, achievement, and attitudes. The survey confirms Japan's very high rates of both achievement and participation in advanced science. Other analyses by Keeves show the importance of rigorous curriculum exposure; given no opportunity to learn, students generally won't. In addition, some now familiar factors influence learning within countries, particularly home environment, prior learning, and student attitudes. Some alterable factors, including teacher competence and an emphasis on inquiry, also will have a bearing on learning. In order to improve science education, Keeves recommends that:

- the study of some science should be mandatory at all levels of schooling;
- strong elementary-school science courses should be provided;
- adequate curricular time should be devoted to science at all levels;
- all science courses should involve an appropriate practical component;
- positive student attitudes toward science should be fostered.

CONCLUSION

As noted in the Foreword, this book was planned by a task force appointed by the International Academy of Education. This volume follows three other published projects sponsored by the Academy: *The Home Environment and School Learning: Promoting Parental Involvement in the Education of Children* (Kellaghan, Sloane, Alvarez, and Bloom, 1993), *Rethinking the Finance of Post-Compulsory Education* (Eicher and Chevaillier, 1993), and *Monitoring the Standards of Education* (Tuijnman and Postlethwaite, 1994). The Academy's initiative and the insights of the task force, authors, and consultants have resulted in an informative and useful collection of essays on science education.

In addition to the usual audiences of scholars, graduate students, and science teachers, this book also specifically meets the needs of policymakers in government and decisionmakers in science education at the national, state, and district levels. Each chapter synthesizes research in a relatively nontechnical way and concludes with practical suggestions for improving science teaching and learning. In this way, it is hoped that this book will fulfill the promise suggested by its title, *Improving Science Education*.

REFERENCES

AAAS (American Association for the Advancement of Science). *Project 2061: Science for All Americans*. Washington, D.C.: AAAS, 1989.

AAAS (American Association for the Advancement of Science). *Benchmarks for Science Literacy* (Project 2061). New York: Oxford University Press, 1993.

Cuban, Larry. "School Reform by Remote Control: SB813 in California," *Phi Delta Kappan* 66 (1984): 213-215.

Eicher, Jean-Claude, and Chevaillier, Thierry. "Rethinking the Finance of Post-Compulsory Education," *International Journal of Educational Research* 19, no. 5, 1993: 449-519 (whole issue).

Fraser, Barry J. "Australia's National Key Center for School Science and Mathematics: Aims and Activities." Paper presented at the annual meeting of the American Educational Research Association, New Orleans, LA, 1994.

Gabel, Dorothy, ed. *Handbook of Research on Science Teaching and Learning.* New York: Macmillan, 1994.

Hurd, Paul D. "Postmodern Science and the Responsible Citizen in a Democracy." Paper presented at the International Conference on the Public Understanding of Science and Technology. Chicago, 1993a.

Hurd, Paul D. "Comment on Science Education Research: A Crisis of Confidence," *Journal of Research in Science Teaching* 30 (1993b): 1009-1011.

Hurd, Paul D. "New Minds for a Modern Age: Prologue to Modernizing the Science Curriculum," *Science Education* 78 (1994): 103-106.

Kellaghan, Thomas; Sloane, Kathryn; Alvarez, Benjamin; and Bloom, Benjamin S. *The Home Environment and School Learning: Promoting Parental Involvement in the Education of Children.* San Francisco: Jossey-Bass, 1993.

National Center for Improving Science Education. *Science and Technology Education for the Elementary Years: Frameworks for Curriculum and Instruction.* Andover, Mass.: The Network, 1989.

National Commission on Excellence in Education. *A Nation at Risk: The Imperative for Educational Reform.* Washington, D.C.: U.S. Department of Education, 1983.

National Research Council. *Fulfilling the Promise: Biology Education in the Nation's Schools.* Washington, D.C.: National Academy Press, 1990.

National Science Board Commission on Precollege Education in Mathematics, Science, and Technology. *Educating Americans for the 21st Century: A Plan of Action for Improving Mathematics, Science and Technology Education for all American Elementary and Science Students so that Their Achievement is the Best in the World by 1991.* Washington, D.C.: National Science Foundation, 1983.

National Science Council. *National Science Education Standards: Discussion Summary.* Washington, D.C.: National Science Council, 1994.

National Science Teachers Association. *Essential Changes in Secondary Science: Scope, Sequence, and Coordination.* Washington, D.C.: NSTA, 1989.

National Science Teachers Association. *Scope, Sequence and Coordination of Secondary School Science, Volume I: The Content Core. A Guide for Curriculum Designers.* Washington, D.C.: NSTA, 1992.

Odden, Allan R., and Marsh, David D. *How State Education Reform Can Improve Secondary Schools, Volumes I and II.* Berkeley: Center for Policy Analysis for California Education, University of California, 1987.

Project 2061. "SFAA Serves as 'Starting Point' for States," *2061 Today* 2, no. 1, 1992: 1-3.

Rutherford, F. James, and Ahlgren, Andrew. *Science for All Americans.* New York: Oxford University Press, 1990.

Task Force for Improving Mathematics, Science, and Computer Education. *A Comprehensive Plan: Improving Mathematics, Science, and Computer Education in*

Florida. Tallahassee: Florida Chamber of Commerce and Florida Department of Education, 1989.

Task Force on Women, Minorities and the Handicapped in Science and Technology. *Changing America: The New Face of Science and Engineering*. Washington, D.C.: National Science Foundation, 1989.

Tobin, Kenneth; and Fraser, Barry J., eds. *International Handbook of Science Education*. Dordrecht, The Netherlands: Kluwer, in preparation.

Tobin, Kenneth; Kahle, Jane B.; and Fraser, Barry J., eds. *Windows into Science Classrooms: Problems Associated with Higher-Level Learning in Science*. London: Falmer Press, 1990.

Tuijnman, Albert C., and Postlethwaite, T. Neville. *Monitoring the Standards of Education*. Oxford, England: Pergamon Press, 1994.

Chapter 2

SCIENCE CURRICULA IN A CHANGING WORLD

John P. Keeves and Glen S. Aikenhead

The importance of science in the school curriculum of most countries continues to grow. Not only do nearly all countries now teach science in some form during the elementary school years (Mayor, 1991), but science also is being taught in the schools of many countries from the first grade. Furthermore, during the years of secondary schooling, science has increasingly become a mandatory subject, with some countries requiring that a science course be taken each year up to the twelfth or terminal grade. Under these circumstances, each country, each school district, and each school at regular intervals needs to reexamine its science program and consider ways in which the science curriculum and the content of its science courses might be improved.

In this chapter we review the historical growth of science education over the previous century, and place developments in science curricula during recent decades in international context. We consider findings from cross-national research studies regarding participation rates, conditions for learning science, and emphasis on different content areas and process objectives across various countries. In addition, we provide some indication of how the teaching of science is responding, and might respond during future decades, to advances in scientific knowledge, to technological change, and to the emergence of critical societal and environmental issues (especially Science-Technology-Society [STS]).

Chapter consultants: Joan Solomon (Oxford University, England) and Sheila Haggis (UNESCO, France).

THE GROWTH OF SCIENCE TEACHING

The development of the formal teaching of science at the school level took place in most countries mainly during the twentieth century. As a consequence, it is possible to trace the common origins and growth of the school science curriculum. The high level of agreement across countries in science curricula makes cross-national comparisons possible, and permits consideration of how science curricula might be improved to meet changing needs and circumstances in countries across the world.

Science was taught in some schools in Britain and the United States during the first half of the nineteenth century. But during the 1860s the establishment in Germany of schools that specialized in the teaching of science, and were separate from those which concentrated on classical studies, laid the foundations for the widespread teaching and systematic learning of science in schools, as well as for the recognition that science had a clearly identifiable place in the secondary school curriculum (Jenkins, 1985). Moreover, the dominance of Germany in all fields of scientific research during the latter part of the nineteenth century led to German texts and programs for science teaching that became widespread throughout Europe, the British Empire, and the USA. In addition, the textbook, the lecture, and the demonstration experiment were adopted in Germany and Western Europe as the appropriate methods by which science should be taught at the school and university levels.

In the USA, the passing of the first Morrill Act in 1862, which created the "land grant colleges" to provide instruction in mechanical arts, military science, and agriculture, helped to establish the teaching of science with an applied emphasis in that country. The importance of practical work and experimentation also was stressed in America. The Harvard list of forty experiments in physics, issued in 1886, and a later statement on laboratory experiments in chemistry, consolidated the emphasis on experimentation. In 1893, the National Educational Association's Committee of Ten recognized that the teaching of science in high schools was an important part of general education and was not merely preparation for university or college. At the elementary school level, a course in nature study was advocated and, at the high school level, year-long courses in geology and meteorology, biology, chemistry, and physics were recommended. These proposals, made approximately 100 years ago, established the particular pattern of science teaching that exists in the USA even to this day.

By the beginning of the twentieth century, laboratory-based science education had become established in many English schools (Gee and Clarkson, 1992), and Armstrong, an English chemist, was advocating the teaching of scientific methods of inquiry through extensive investigatory work (Armstrong, 1903). The acceptance of the central place of practical work in the teaching of science required that a heavy commitment of time be given to science subjects, with a marked degree of specialization by some students in science in contrast to the humanities. The science teaching that resulted in England was characterized by (a) an early and extreme specialization in science, (b) the use of a concyclic method of presentation of content (in contrast to the alternative method which had developed in the USA), and (c) a marked emphasis on laboratory training.

While the content of science courses at the university level and subsequently at the school level in the USA and England was derived from courses provided in Germany, each of these countries developed its own particular orientation toward the teaching of science and its presentation to secondary school students. Moreover, each of these three countries had a marked influence on the science courses that were established in other lands. In many parts of the world, such as Australia, the influence of all three countries can be found at different times over the past 100 years.

THE COMMON BODY OF SCIENCE CONTENT

The specification of the content of the science curricula taught in schools across very different cultures is dependent on the existence of a clearly identified body of basic scientific ideas and principles. These are agreed upon widely and provide the foundations for understanding the natural world, for advancing technological development, and for undertaking the quest for scientific knowledge. Moreover, there is a widely accepted set of tactics and strategies of science, to use Conant's (1947) phrase, that provides the framework within which scientific inquiry is conducted. Without such agreement, science could not have advanced as it has during recent times.

Moreover, a detailed examination of the science curricula of those school systems which participated in the Second IEA (International

Association for the Evaluation of Educational Achievement) Science
Study in 1983-1984 revealed that there is a largely common body of scientific content across countries that is taught to students with a high
level of emphasis at the terminal secondary school stage in the fields of
biology, chemistry, and physics (Rosier and Keeves, 1991). Countries can
differ markedly in how their science courses are organized during the
twelve years of schooling. They also can differ in the emphases that they
place on laboratory work and investigation. Furthermore, they can differ
to a marked degree in the proportions of the age groups that they retain
at school at different grade levels and, in particular, at the terminal stage
of schooling. Nevertheless, there is little difference among countries in
the emphases placed on topics in the fields of biology, chemistry, and
physics that are taught to science specialist students before they proceed
to university or higher education. This should not be taken to imply that
all countries reach an equally high standard of achievement. (The reasons for differences in levels of achievement between countries are considered in chapter 10 of this volume.) However, there is a large measure
of common agreement across the world on the content of science curricula that is appropriate during the years of secondary schooling in the
three major fields of biology, chemistry, and physics.

The reasons for this convergence in the content of science courses
are twofold. First, there is universal agreement among scientists with
respect to the foundational body of content on which scientific investigation has built during the past 100 years. Nevertheless, this body of content in the fields of biology, chemistry, and physics has not remained
unchanged during this period. Indeed, it has developed rapidly and it
would appear to be growing at an increasing rate.

Second, the establishment of school science programs—drawing on
the methods and content of biology, chemistry, and physics in university
courses, initially derived from Germany and subsequently in some countries from England and the USA—has resulted in a high level of agreement about what should be taught in schools to provide the foundational
knowledge for further study in these fields. This high degree of consensus does not occur in other subject areas of the school curriculum when
examined cross-nationally. Even in mathematics, there are very different
traditions in the content of courses that are taught in countries that have
been influenced by English and French educational programs.

THE OTHER FIELDS OF SCHOOL SCIENCE

In the field of earth science, there are very different traditions and emphases in different parts of the world. In continental Europe, the subject of geography or earth science was oriented initially toward geology, and was considered a science. However, in recent decades in those countries, this subject at all levels has been directed toward social and economic geography, and has tended to be oriented more strongly toward the social sciences. As a consequence, in school science courses in most countries, only slight or moderate emphasis is placed on topics in the field of earth science and geology, with many related topics being taught in geography courses. Furthermore, at the terminal secondary school level, only in four of the twenty-three countries involved in the Second IEA Science Study (Australia, Italy, Japan, and South Korea) is earth science formally taught. Three of the four countries that teach earth science at the terminal secondary school level lie within active volcanic zones.

In some countries, although rarely among those of the western world, topics from the behavioral sciences have been incorporated into the science courses taught in schools. Whether this move will continue to grow remains to be seen. A more likely field of development in science education is the inclusion of the subject of technology into the courses provided for all students, either integrated with other areas of science or taught as a separate subject, as is the practice in Swedish lower and middle secondary schools. However, since the 1970s, in some countries (England, for example) courses in technology have been provided for some students at certain grade levels. The Second IEA Science Study report by Rosier and Keeves (1991) presents more detailed information about the teaching of science in schools in twenty-three countries in the mid-1980s.

While there is not complete and universal agreement across countries with respect to the content of school science courses at successive levels of schooling, there is general agreement among countries about the content objectives in the fields of biology, chemistry, and physics at the terminal secondary school level. This high level of agreement has made the study of science achievement across countries, as carried out by the International Association for the Evaluation of Educational Achievement, a worthwhile enterprise.

SCIENCE CURRICULUM REFORM IN THE MID-TWENTIETH CENTURY

In the decade after the conclusion of the Second World War, the efforts of scientists, technologists, and educators went into rebuilding countries that had been ravaged by the war. Associated with this rebuilding was the reshaping and expansion of secondary education. However, by the mid-1950s in both Great Britain and the USA, there was a growing realization that school science courses also needed to be improved to take into account new understandings of the natural world achieved by science as well as new perceptions of the processes of scientific inquiry. In both countries, this work of redevelopment of school science courses had been initiated prior to the launching of Sputnik by the USSR in 1957, but the very obvious advances of Russian science and technology were used to focus public attention, particularly in the USA, on the urgency of reform in science education.

The movement for science curriculum reform spread quite rapidly around the world during a twenty-year period until the mid-1970s, when it lost momentum (Welch, 1979). While it can be argued that the new well-packaged science courses developed by major curriculum projects were expensive to introduce, and did not achieve widespread adoption (Walberg, 1991), there is little doubt that the reform movement itself had a profound and lasting effect on the teaching of science across the world. Among the common features of this reform movement in many countries are the following:

1. *Modernization of content in chemistry and physics.* The marked advances achieved in the fields of chemistry and physics during the first half of the twentieth century were incorporated as basic ideas and principles into the teaching of these subjects. This required the elimination of quite large segments from the science courses taught in the middle and upper secondary schools. In physics, for example, a greater emphasis was placed on waves, and the emphasis on introductory heat, sound, and electrostatics was reduced. In chemistry, the emphasis on inorganic chemistry was decreased and greater emphasis was placed on theoretical aspects concerned with the nature of the chemical bond.

2. *Emergence of biology as a coherent field.* During the first half of the twentieth century, biological topics were taught in a fragmented way

either as botany, zoology, and physiology, or as subordinate to chemistry and physics. The reform movement emphasized that the field of biology existed in its own right, and integrated the different components around ecological principles, the nature and function of the cell, or the functions of the living organism. One consequence of the redevelopment of biology teaching has been the increased numbers of students taking courses in biology at the upper secondary level in some countries.

3. *Renewed emphasis on investigation and laboratory work.* Up to the mid-1950s, the texts used in science teaching were largely descriptive. Science was presented almost wholly as a collection of loosely connected facts and elementary generalizations. The reform movement sought to show science as an experimental enterprise, with students developing some of the skills of scientific inquiry and obtaining at least some of their knowledge through investigations they carried out for themselves.

4. *Introduction of elementary school science.* Science in the elementary school traditionally had been taught largely as nature study. This approach had the advantage that students were encouraged to learn through careful observation and classification, but it ignored much of the natural environment that had an impact on students' lives. Gradually, the teaching of science (biology, earth science, chemistry, and physics) has been introduced at the elementary school level in most countries in a way that links together ideas from all fields and relates them to the students' immediate surroundings and everyday experiences.

5. *Reduced emphasis on the applications of science.* In many countries, but perhaps not all, one of the consequences of the upgrading of the content of science courses in the secondary school was the increased emphasis on theoretical ideas and principles and the reduced emphasis on the applications of science both in industry and everyday life. In part, this was a necessary and desirable reaction to the practice that had developed in the USA during the second quarter of the twentieth century of presenting a mass of poorly connected facts and of relating them to the gadgets and devices that modern manufacturing had provided for the home and the workplace. Thus, curriculum reform had the effect of divorcing the science taught in schools from technology and its applications. The

reforms also gave rise to science courses that were more theoretical and abstract and had little appeal to anyone but the most able students. Moreover, the new courses provided a distorted view of the relationships that existed between science, technology, and society, and thus an inadequate understanding of the role of science in the modern world.

There is little doubt that the teaching of science around the world profited greatly from these curriculum reforms. However, they placed heavy demands on teachers and were rarely introduced in ways that were originally planned and advocated. Moreover, the increased emphasis on investigation and laboratory work required that schools be provided with appropriate laboratory facilities and equipment. Although programs were introduced in some countries (for example, Australia and England) in an attempt to remedy such deficiencies, in less affluent countries and school districts the lack of appropriate facilities prevented these reforms from being introduced effectively. This would appear to be particularly true in the USA.

MAPPING THE SCIENCE CURRICULUM IN THE MID-1980s

As already mentioned, the wave of science curriculum reform came to an end in the late 1970s and, before planning the next phase of science curriculum development, it was considered desirable to map the state of science education across the world. This led the International Association for the Evaluation of Educational Achievement (IEA) to undertake the Second IEA Science Study (SISS) in 1983-1984. Furthermore, the opportunity was provided for changes to be examined over the fourteen-year period following 1970-1971 when the First IEA Science Study was carried out. In the section that follows, evidence concerning each of the five reforms considered above is examined by drawing on the findings of the SISS. For further information about these findings, the reports of this study by Rosier and Keeves (1991), Postlethwaite and Wiley (1991), Keeves (1992a), and Keeves (1992b) could be consulted. The presentation of evidence collected in the SISS project is limited to twenty-four school systems from twenty-three countries, to only one representative (Canada [Quebec]) from the French-speaking countries, to none of the Spanish-speaking and Arabic-speaking countries, and to only limited representation from Eastern Europe, Africa, and Asia. Nevertheless, it is the only study for which detailed comparative data are available over time. However, this does not invalidate the identification of general trends that are highly consistent across the countries examined.

In planning the SISS project thought was given to new science content that might have been included in the curricula of many countries between 1970 and 1984. Three general areas were considered to be highly important in this regard in the developed countries: (a) the history and philosophy of science; (b) environmental science; and (c) technical and engineering science. In addition, SISS sought in 1984 to include a greater number of developing countries than were included in the 1970 study. It was argued that in these countries greater emphasis was likely to be given to two additional content areas: (d) rural science; and (e) health science. Consequently, it was decided to collect information about the degree of emphasis in the science curricula of the countries engaged in the second study regarding the following five content areas:

History and Philosophy of Science: historical development of science, nature of science

Environmental Science: energy resources, energy use, environmental impact, habitats

Technical and Engineering Science: transport, manufacturing processes, computers and microprocessors

Rural Science: animal husbandry, plant husbandry, housing and rural amenities

Health Science: personal health, interpersonal relationships, community health

Curriculum experts were asked to rate on a four-point scale the level of emphasis given to each of these topics. It was considered that within each content area the same topics could be treated at different levels of conceptual difficulty at the different stages of schooling. The ratings were averaged across countries and separately for the three student populations involved in the SISS (10-year-olds, 14-year-olds, terminal secondary school level).

There was little evidence of a concyclic development of these topics in the science curriculum across the stages of schooling. Only limited emphasis, other than for the topics of environmental impact and habitats, was placed on these topics at any level of schooling. Nevertheless, associated with each of the topics are issues that need to be addressed at some stage in the teaching of science during the twelve years of schooling. While the content areas of Rural Science and Health Science are of particular importance in developing countries, some consideration might be given to the issues raised in these areas in the science courses of the developed and more industrialized countries. Further details can be found in Rosier and Keeves (1991).

Participation in Science Courses

In some countries, the reforms introduced during the twenty-year period from the late 1950s to the late 1970s sought, among other goals, to increase participation in the study of science at the terminal secondary school level. Such participation is influenced, in part, by the proportion of the age group remaining at school to the terminal secondary school level.

In the decades following the end of the Second World War, there was a marked growth in secondary school enrollments throughout the western world. This arose partly from the natural increase in population of many countries. However, it also came from the striking trend to stay in school longer. During the years between the undertaking of the first and second IEA science studies, this trend to stay longer at school continued at the terminal secondary school level. Table 2-1 gives the proportions of the age group enrolled on two occasions, namely, 1970-1971 and 1983-1984, at the terminal grade level (which was Grade 13 in England and Italy and Grade 12 in all other countries). Only in Sweden was a decline in participation recorded: from 30 percent of the age group in 1970-1971 to 28 percent of the age group in 1983-1984. In England, the proportion of the age group at school at Grade 13 remained constant over time at 20 percent, although an additional 4 percent of the age group was enrolled in colleges of further education in 1983-1984. Of those students enrolled at the terminal secondary school level, sizeable proportions were in vocational schools in Finland (14 percent), Hungary (22 percent), Japan (26 percent), and Thailand (15 percent). These schools, in general, did not prepare students for entry to a university or college and were not included in the target populations for the SISS project.

While the proportion of the age group forming the science specialist group grew in some countries as a result of the trend to stay longer at

Table 2-1

Changes in Participation in Science Courses at the Terminal Secondary School Level Between 1970-1971 and 1983-1984

Statistic	Year(s)	Enrolment and Participation Rates								
		Australia	England	Finland	Hungary	Italy	Japan	Sweden	Thailand	USA
Enrolled at the Terminal Grade										
% of Age Group	1970	29	20	21	28	16	70	30	10	75
	1983	39	20	59	40	34	89	28	29	80e
% of Group Tested	1983-84	39	20	41	18	34	63	28	14	80e
Science Specialists										
% of Age Group	1970-71	18	9	16	22	10	-	16	9	54
	1983-84	29	10	14	8	13	28	13	7	21
Participation Rates (% of Age Group)										
Biology	1970-71	9	4	14	17	1	-	9	9	12
	1983-84	18	4	41	3	4	12	5	7	6
Chemistry	1970-71	11	5	5	2	1	-	8	8	13
	1983-84	12	5	16	1	1	16	6	7	2
Physics	1970-71	11	6	9	22	8	-	14	8	20
	1983-84	11	6	14	4	13	11	13	7	13
Non-Science Specialists										
% of Age Group	1970-71	11	11	5	6	6	-	14	1	19
	1983-84	10	10	27	10	21	35	15	7	49
% of Non-Science Specialists not Taking Any Science	1970-71	51	100	100	100	100	-	100	10	96
	1983-84	73	100	0	0	22	100	100	21	100
% of Age Group not Taking Any Science	1970-71	6	11	5	6	6	-	14	0.1	18
	1983-84	7	10	0	0	4	35	15	1	49

e Estimated value

school, in Sweden, Thailand, and the USA the proportions fell. This was noticeably so in the USA. The increased retention rates through staying longer at school also led to growth in six countries in the proportion of the age group which comprised the non-science specialist students. In 1970-1971, the non-science specialist students, in general, were not taking any science courses. However, by 1983-1984, all students in Finland and Hungary at the Grade 12 level were required to study some science, as were more than three-quarters of the students at the terminal secondary school level in Italy.

This raises major questions about the type of science appropriate for students who do not intend to specialize in science. Furthermore, it raises the issue as to whether the study of some science should be made mandatory throughout the twelve years of schooling. It can be seen from table 2-1 that sizeable proportions of the age group were in school in academically oriented courses at the Grade 12 level, but were not required to study any science at the time of testing in 1983-1984 in Japan (35 percent) and the USA (49 percent). There were also significant proportions of the age group enrolled at school but not studying science at the terminal secondary school level in Australia (7 percent), England (10 percent), Italy (4 percent), and Sweden (15 percent).

Of particular interest is the change in the proportion of the age group that was enrolled in biology courses at the terminal secondary school level. In the absence of data to indicate the situation in the mid-1950s before reforms were introduced, it is difficult to assess the effects of the changes that occurred through the modernization and restructuring of science courses. However, the data presented in table 2-1 indicate that only in three of the eight countries for which comparisons can be made was there an increase between 1970-71 and 1983-84 in the proportion of the age group studying biology. In Australia, under circumstances in which there was a marked increase in retention rates, the gains also were the result of major reforms in the teaching of biology. Likewise, in Finland and Italy there also were marked increases in retention rates between 1970-1971 and 1983-1984, with biology being a mandatory school subject in Finland. However, in two of the remaining countries in which there were increases in retention rates, namely, Thailand and the USA, there were declines in the proportion of the age cohort studying biology. Moreover, in England and Sweden, where the retention rates showed little change between 1970-1971 and 1983-1984, there was little change in the proportion of students studying biology in England and there was a clearly recognizable decline in Sweden. Thus, the science curriculum reforms did not always have the effect of maintaining participation in the study of particular fields of science.

Teaching the Investigatory Processes of Science

The learning of science involves not only the mastery of ideas and principles. It also involves the acquisition of the skills of inquiry. These skills are learned primarily in the school science laboratory and through practical work. The teaching of these skills was examined by asking the countries participating in the second science study to provide ratings of the emphasis given to this aspect of the science curriculum. The following process objectives of science as inquiry were identified, using the work of Klopfer (1971), which built on ideas developed in the first science study:

Processes: knowledge and understanding, observation, measurement, problem solving, interpretation of data, formulation of generalizations, model building
Applications of Science
Manual Skills
Attitudes, Interest, and Values
Limitations of Science and Scientific Methods

The seven objectives related to the processes of science were considered to be ordered hierarchically. The ratings used were 3 (major emphasis), 2 (minor emphasis), 1 (low emphasis), and 0 (not included in the curriculum). Seventeen of the twenty-four education systems involved in the study were able to provide ratings. In the remaining seven systems, there would appear to have been little emphasis on these process objectives in the science curriculum.

Figure 2-1 gives the mean ratings averaged across countries for each population level for those countries that responded to the questions about emphasis on science processes. Generally, emphasis on process objectives increases as students progress through the stages of schooling. Moreover, there was a strong emphasis on knowledge, observation, measurement, problem-solving, the interpretation of data, manual skills, and attitudes, interests, and values. However, there is clearly a low level of emphasis, even at the terminal secondary school stage, on the formulation of generalizations, model building, and the limitations of science and scientific methods. During their final years of schooling, these students learn and develop skills in working with abstract concepts and logical relations, frequently through the use of mathematical procedures; little emphasis, however, is placed on the development of an understanding of the nature of scientific knowledge and its growth during recent centuries throughout the world.

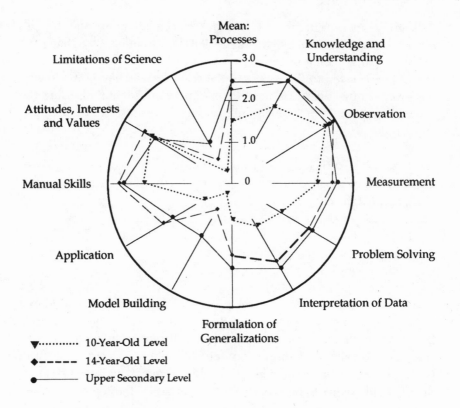

Note: *Radial length indicates degree of emphasis on scale 0 to 3.*

Figure 2-1

**Emphasis on Process Objectives in Science Curricula for the
10-year-old, 14-year-old, and Terminal Secondary School Levels**

The Applications of Science

Figure 2-1 also documents an important result that emerged from the reforms in science teaching which were introduced during the twenty-year period from the late 1950s to the late 1970s. It was noted above that one consequence of the upgrading of the content of science courses to take into account the emergence of new theoretical perspectives in the fields of biology, chemistry, and physics was that the emphasis placed on the applications of science was reduced greatly in most countries. When data are averaged across countries a lower level of emphasis on application than on other process objectives is clearly evident at all three stages of schooling, as shown in figure 2-1, thus indicating that relatively few countries are exceptions to this generalization. In addition, there is less emphasis on application at the science specialist level at the terminal stage of secondary schooling than at the 14-year-old level. This reduced emphasis on topics concerned with the applications of science also is evident in the more detailed curriculum analyses carried out by topic. Those topics concerned with electronics, machines, the chemistry of life processes, environmental chemistry, chemistry in industry, and soil science, on average across countries, have a much lower level of emphasis accorded to them at the terminal secondary school stage than do the other more theoretical topics (Keeves, 1992a; Rosier and Keeves, 1991). It is not possible from the evidence available, however, to establish a decline in emphases in science curricula over time on the applications of scientific knowledge to practical industrial situations. However, it is important to note that many ex-colonial countries have engaged in a slow struggle to make their science courses more relevant to national needs through an emphasis on the application of scientific principles to local problems.

The Introduction of Science into the Elementary School

One of the major changes that resulted from the reform movement in science teaching was an increased emphasis on the teaching of science in the elementary school. Based on information available from the data collected in the two IEA studies, table 2-2 provides details about (a) the age at which the learning of science begins; (b) the grade level at which learning science begins; (c) the percentage of students reporting the use of a textbook in science lessons; and (d) the percentage of students reporting that they undertake experiments in science lessons.

The overall view obtained from these data is one of an increased emphasis on science teaching over the period from 1970-1971 to 1983-1984. With the exception of Sweden, where it was possible for students to start learning science at a younger age or an earlier grade, a change was made. However, in Australia and Italy, not all schools sampled were teaching science in the early grades, although a majority were doing so. The evidence presented also indicates a marked increase in the propor-

Table 2-2

**Conditions of Learning Science at the 10-Year-Old Level,
1970-1971 and 1983-1984**

Country	Age Beginning Science		Grade Beginning Science		Use Textbook in Science (%)		Do Experiments in Science (%)	
	FISS	SISS	FISS	SISS	FISS	SISS	FISS	SISS
Australia	8	6	3-4	1	na	45	na	72
England	9	5	4	1	45	62	21	82
Finland	7	7	1	1	74	97	10	61
Hungary	6	6	1	1	74	99	28	63
Italy	8	v	3	v	58	91	23	63
Japan	6	6	1	1	100	97	86	92
The Netherlands	8	v	2	v	59	na	46	na
Sweden	10	10	4	4	86	95	47	58
Thailand	8	7	1	1	78	na	73	na
USA	6	6	1	1	74	90	47	77

na   Data not available   FISS   First IEA Science Study: 1970-1971
v   Various   SISS   Second IEA Science Study: 1983-1984

tion of students reporting that they used a textbook in science lessons. In addition, the data show very substantial increases in the proportions of students reporting that they did experiments in science lessons. These two aspects of science teaching in elementary school classrooms suggest that, over time, there was a general and substantial increase in the extent to which science was taught in a formal and structured way.

EDUCATION IN A WORLD OF CHANGE

In recent decades, not only have science and technology made spectacular advances, but the societies in which we live, work, and learn have undergone some dramatic changes. While many scientific and societal changes might be identified, it is sufficient here to focus on five changes that have important consequences for education, particularly for the teaching of science in schools.

1. *Toward universal secondary education.* In the forty-year period following the end of the Second World War, there was a quite dramatic expansion of secondary education in many countries. This led to a gradual shift from highly selective academic schooling to the provision of a more general education for steadily increasing proportions of the age cohort. Many countries thus are moving rapidly toward a situation in which universal secondary education is provided, with a very high proportion of the age group remaining at school for up to twelve years. Such changes demand a reexamination of the content of science courses offered during the later years of secondary schooling, with some repercussions about what should be taught at the earlier stages.

2. *Lifelong education.* There has emerged a recognition of the need for adults to undertake continuing education and repeated retraining throughout their lives. This need arises from the rapid growth of knowledge, from the demand for adults to understand the changes which are occurring in society, and from the requirement for the work force to acquire new skills as the operations of industry and commerce undergo rapid change. The provision of life-long education programs is shared among universities, colleges of further education, industry and commerce, and the trade unions, and is becoming the major field of educational growth. However, more important than the organizational arrangements made to provide programs of continuing education are the consequences for what is taught in secondary schools and how learning occurs. It is clear that, under the circumstances in which education takes place at intermittent stages throughout a lifetime, it is necessary for schools to provide students with the foundational knowledge on which they can build at a later time. This is particularly important in the field of science because further growth of scientific knowledge and understanding is likely to be required during adult life. Thus, on leaving high school, all students should have a sound basic knowledge of science in the fields of earth science, biology, chemistry, and physics.

3. *Learning to learn.* One of the consequences of the rapid growth of scientific knowledge is the need for each individual to acquire the skills of effective independent learning and inquiry. The procedures employed in scientific investigation are key skills for learning. Consequently, all students graduating from high school should have mastered these skills of inquiry as part of learning how to learn. To achieve this

goal, the investigatory aspect of science teaching needs to receive
greater emphasis, and there must be a shift away from the performing of
routine experiments (involving the following of standardized instructions) that are common in a traditional school science laboratory.

4. *Emergence of science-related social issues*. During the 1980s,
while the SISS was in progress, the magnitude of the many problems
facing all countries and all people scattered across the "global village"
became clear. These problems included:

- the population explosion;
- the emergence of new drug resistant diseases;
- the feeding of starving people in Third World countries;
- the effects of genetic experimentation and engineering;
- the ecological impact of modern technology;
- the environmental effects of deforestation of jungle and forest regions;
- the desertification and degradation of land in arid and semi-arid zones;
- the dangers of nuclear war and explosions of nuclear installations; and
- the possibility of major climatic change.

The mass media have publicized these problems and have promoted
debate. As a consequence, students, particularly at the upper secondary
level, not only are aware of such problems but they seek the knowledge
to debate meaningfully the resolution of the issues raised. Science
courses in schools no longer can afford to ignore such problems, lest the
courses are seen as irrelevant by students. Furthermore, without an
informed public, there is little chance that resources would be made
available for the resolution of these problems. Science education has to
accept responsibility for control instead of catastrophe in human affairs.
In addition, science education has to be seen as lifelong and recurrent,
and not as restricted to the stages of elementary and secondary schooling,
because new issues undoubtedly will emerge during coming decades.

5. *The impact of technological change*. In the late 1970s and early
1980s, it became apparent that: (a) advances in microelectronics would
change the nature of employment in the office and the factory; (b) new information technology not only would influence the storage and retrieval of
information in libraries, but it would also modify greatly the procedures

employed in commerce and industry; (c) new technology had the potential to change modes of transport and the sources of energy for industry and the home; and (d) biological technology would change the production of food and beverages, the provision of health and medical services, and plant and animal breeding. Such widespread and radical changes demanded a greater knowledge of scientific principles by three groups of people. First, there were those who would be employed in science-based professional occupations. Second, there were those who would be employed as highly skilled workers in industry. Third, there were all those people in the general population who needed to understand and debate within a democratic society the issues raised by scientific development. As a consequence, there has been further demand for curriculum reform in science. However, the strategies of reform that were employed two decades earlier no longer are seen to be appropriate. Moreover, there is no widely accepted view of the directions which such reforms should follow. No longer is the primary goal of reform the upgrading of scientific content.

THE EMERGING RESPONSES OF SCIENCE EDUCATION

Science education is responding to the changes discussed above in several clearly identified ways, although these responses as yet have not gained widespread acceptance, even within a particular country or school system, let alone across a group of countries. Five areas have arisen in which responses are required:

1. consideration of environmental issues, which has led to the introduction of environmental education courses;
2. growth of the Science, Technology, and Society (STS) movement, which has led to a wide range of STS courses, many of which address the science-related social issues listed above (with these courses sometimes focusing on environmental questions, thus leading to the development of science, technology, and environment [STE] courses);
3. interest in the history, philosophy, and sociology of science, which commonly has been incorporated into courses in philosophy and the nature of knowledge;

4. establishment of technology as a mandatory subject in the curriculum, which involves recognition that the specific skills of woodwork and metalwork for boys, and needlework for girls, are becoming obsolete, and that more general skills of problem solving and design are required as well as skills that are oriented toward computerized technology;

5. emphasis on investigation in science, which has led to the reorientation of practical work toward investigatory projects, the learning of the process skills involved in scientific inquiry, and the learning of how to search for information and assemble data using computer-based procedures.

The demands made on science education are not only diverse but also immense. Further, these demands exist both in the advanced technological societies and in the developing countries of Africa and Asia. While some problems could seem more pressing in certain countries than in others, none can be ignored safely by any one country. Such is the nature of our interdependence on this planet. Nevertheless, it must be argued that the prime tasks of science education lie in the dissemination of scientific knowledge, the development of the ability to apply scientific principles, and the inculcation of skills associated with the processes of science as a method of inquiry. However, the production of a generation of students who are very knowledgeable in science, but who lack an understanding of the problems listed above, is to invite disaster. Likewise, to produce a generation of students who are aware of issues but who lack competence in science, and are ignorant of the scientific principles which might be applied to these problems, could be equally disastrous. Some programs and approaches developed within the STS movement are considered in the next section.

THE SCIENCE, TECHNOLOGY, AND SOCIETY MOVEMENT

Widespread recognition of the interrelations among science, technology, and society came from the publication of the work *On Understanding Science* by James B. Conant (1947), then President of Harvard University. The importance of this publication was recognized quickly and it was used in some schools as a basis for the teaching of science to non-science specialists in the early 1950s. Subsequently, Klopfer and Cooley (1963) prepared a series of twelve booklets on *History of Science Cases for High School*, which provided a detailed account of the relationships among science, technology, and society for twelve important scientific developments that had involved major advances in scientific theories. Cooley and Klopfer (1961) also prepared the *Test On Understanding Science* (*TOUS*) to measure students' understanding of the nature of science and scientific discovery. Again, from Harvard University came the *Harvard Project Physics* course (Holton, Rutherford, and Watson, 1970), which reorganized the teaching of high school physics so that the historical and conceptual development of the subject in relation to technology and society was recognized more clearly. From the mid-1970s, the field of science, technology, and society, with an orientation toward social issues, has emerged as one of growing importance in school syllabuses (Solomon, 1990). This development is far from complete, but it has forced a reexamination of how science is taught at different levels of schooling. Moreover, it is likely to play a critical role in the reshaping of science education in the decades ahead.

Ziman (1980) has drawn attention to a number of different models of science, and recognition of these models is necessary if the teaching of science in schools is to be placed in an appropriate context. Three of Ziman's models are of interest. The first view of science is as an activity in which human beings struggle to produce knowledge about the natural world. Links with society's goals and beliefs are rejected and the knowledge gained is a reflection of the physical and biological worlds as they exist.

The second view of science is as an instrument of national purpose. In this instrumental model, science advances technological development, promotes national wealth, fosters improved health, and provides more effective forms of warfare. Science itself is a black box that need not be understood or examined by the public at large or even by scientists.

In the third view of science, the interrelations among science, technology, and society are appreciated. It is recognized that the scientific community constructs the knowledge that it generates, and verifies its constructions against the natural world. Scientific knowledge is endorsed through a web of belief and it is confirmed by the usefulness of that knowledge to explain and predict. It is this third model of science that justifies the increased commitment to science in the school curriculum. This is because the role of science in schools is not merely to train a cadre of research scientists, nor merely to improve the technological skills of the workforce in order to advance national prosperity, although both of these objectives are important. The primary purpose of teaching science in the schools is to provide students with a way of thinking and inquiry that is the most powerful currently available for everyday living, for scientific research, for fostering the technological and economic growth of the societies in which they live, and thus for improving standards of living.

This development of the skills of investigation and inquiry is incomplete without an understanding of the view of science contained in the third model above. Furthermore, only with this appreciation of science is the high school graduate provided with the understanding and skills necessary to question the application of scientific knowledge in particular situations, to monitor the effects of scientific discovery on the quality of life, and to modify the use of natural and human resources. If high school graduates are not provided with this understanding of science and its methods, they are likely to plunder the natural environment, waste human resources in the quest for immediate pleasure and gratification, and seek to explain and predict events through astrology and mystic belief.

The purposes of teaching science in schools commonly are limited to three (see Skilbeck, 1984). First, science is taught to pass on knowledge to young people who will carry that knowledge forward. Second, science is taught for vocational study. Third, science is taught so that each generation can reconstruct its own society. However, these three purposes, although unquestionably valid, are incomplete and inadequate. Science also is taught in schools as a way of learning the skills of investigation and inquiry. In addition, science provides a way of explaining and predicting events in the natural world that human beings require for effective living. Thus, in spite of the limitations of both scientific knowledge and scientific methods, science affords a very powerful approach to learning which students at school should acquire if they are to continue learning throughout their lives. Unless students are introduced to the ideas presented in science, technology, and society courses, their views of science are limited to the first of both Ziman's (1980) models and Skilbeck's (1984) three purposes. Without such a course, the possibility of social reconstruction is not raised and, above all, the methods of scientific investigation are not seen in full perspective. A science, technology, and society course, of its own, does not develop the skills of inquiry, but it does provide an understanding of how scientific inquiry operates in the modern world. Without this understanding of the processes of scientific inquiry, the high school graduate would not be well-placed to harness the powerful forces of scientific knowledge to improve the lives of fellow human beings.

The Content of STS Courses

A wide variety of views exists as to the most appropriate content for STS courses. Aikenhead (1994) suggests that, from a science education orientation, STS content can be defined by (a) interactions among science, technology, and society and (b) any one of the following: a technological artifact, process, or expertise; a societal issue related to science and technology; social science content that sheds light on a societal issue related to science and technology; a philosophical, historical, or social issue within the scientific community; or any combination of the above. It should be noted that, in this definition, there is concern for two types of social issues as identified by Rosenthal (1989): (a) social issues external to the scientific community (for example, issues involving energy conservation, or pollution) and (b) social issues internal to the scientific community (for example, the nature of scientific theories or the cold fusion controversy).

Associated with this very broad definition of STS content is a diversity of materials that have been developed during the past twenty years and that purport to present STS courses. In the section that follows, a classificatory scheme advanced by Aikenhead (1994) is presented which is concerned more with structure than with educational function or goal.

The Structure of STS Courses

The classificatory scheme shown in table 2-3 characterizes STS courses in terms of (a) the proportion and organization of traditional science content compared with STS content; and (b) the emphasis given to traditional versus STS content in the assessment of student performance. In addition, each category is defined operationally by reference to published texts and courses drawn from different parts of the world

(Australia, Canada, the Netherlands, the UK, and the USA) where STS courses have been developed.

The methods of presentation employed and the logical development of content depend largely on the category level involved. For more STS-oriented courses, a particular sequence likely to be used involves five

Table 2-3

A Classification of STS Courses

No.	Name of Category	Description of Category	Emphasis in Student Assessment	Examples
1	Motivation by STS content	Standard school science is taught, together with mention of STS content to make lessons more interesting.	Students are not assessed on STS content.	*A Second Course in Mechanics*, UK: McKenzie
2	Casual infusion of STS content	Standard school science is taught, together with a short study of STS content attached to the science topic. The STS content does not follow cohesive themes.	Students are assessed mostly on pure science content and only superficially on STS content.	*SATIS*, UK: Association for Science Education; *Values in School Science*, USA: Brinckerhoff, *Consumer Science*, USA: Burgess
3	Purposeful infusion of STS content	Standard school science is taught, together with a series of short studies of STS content integrated into science topics in order to explore systematically the STS content. This content forms cohesive themes.	Students are assessed to some degree on their understanding of the STS content.	*Science and Social Issues*, USA: Welch; *Science and Societal Issues*, USA: Iowa State University; *Science, Technology and Society*, USA: New York State Education
4	Single discipline through STS content	STS content serves as an organizer for the science content and its sequence. The science content is selected from one science discipline.	Students are assessed on their understanding of the STS content, but not to the same degree as on the pure science content.	*Project Physics*, USA: Holt, Rinehart and Winston; *Light Sources*, The Netherlands: PLON; *Science and Society Teaching Units*, Canada: OISE
5	Science through STS content	STS content serves as an organizer of science content and its sequence. The science content is multidisciplinary.	Students are assessed on their understanding of STS content, but not as extensively as they are on the pure science content.	*Logical Reasoning in Science and Technology*, Canada: Wiley; *Modular STS*, USA: Wausau, Wisconsin; *The Dutch Environmental Project*, The Netherlands: NME-VO, University of Utrecht
6	Science along with STS content	STS content is the focus of instruction. Relevant science content enriches the learning.	Students are assessed equally on the STS and pure science content.	*Society, Environment and Energy Development Studies (SEEDS)*, USA: SRA; *Science and Technology II*, Canada: BC Ministry of Education

(Cont.)

Table 2-3

A Classification of STS Courses (continued)

No.	Name of Category	Description of Category	Emphasis in Student Assessment	Examples
7	Infusion of science into STS content	STS content is the focus of instruction. Relevant science content is mentioned but not systematically.	Students are assessed primarily on the STS content, and only partially on pure science content.	*Science in a Social Context*, SISCON, UK: Association for Science Education; *Science, Technology and Society*, Australia: Jacaranda Press; *Modular Courses in Technology*, UK: Schools Council
8	Pure STS content	A major technology or societal issue is studied. Science content is mentioned, but only to indicate an existing link to science.	The students are not assessed on the pure science content to any appreciable degree.	*Science and Society*, UK: Association for Science Education; *Innovations: The Social Consequences of Science and Technology*, USA: BSCS; *Values and Biology*, USA: Welch

Adapted from Aikenhead (1994)

phases: (1) a social question which leads to, (2) pertinent technology that defines, (3) relevant science content that sheds more light on, (4) related technology, and (5) the original social question (Aikenhead, 1990). Moreover, research carried out on the Dutch Physics Curriculum Development Project (known as "PLON" in Dutch) has shown that this sequence of presentation of content is highly effective in fostering among students a need to know and understand the scientific content (Eijkelhof and Lijnse, 1988).

The different category levels at which courses are planned and materials are developed without doubt are related to requirements for entering higher education, to the freedom allowed to schools, and to the extent to which teachers within schools are permitted to select courses of their own choice. The decision to adopt an STS-type course in preference to a traditional science content course is not based primarily on the effectiveness of student learning assessed in terms of traditional science content. Many studies have concluded consistently that both types of science courses are equally effective in terms of students' mastery of traditional science content. Rather, the decision is related more to the aims and objectives of teaching science to a particular group of students and to whether a particular STS-type course meets these objectives better than a traditional course. Increased retention rates at the higher levels of secondary schooling demand a greater variety of science courses to meet a greater diversity of needs of students who have a greater range of interests, abilities, and aptitudes. It is at the upper secondary school level that courses with a greater emphasis on STS content are more likely to be used. Moreover, at all levels of schooling, some emphasis on STS content should be introduced into courses in order to begin to develop in students some initial awareness of the complex relationships that exist among science, technology, and society, as well as an introductory understanding of the complex nature of scientific knowledge. Furthermore, as students move to the upper levels of secondary schooling, they are more likely to become aware of and able to debate the many complex issues related to science that have arisen in modern technological societies.

IMPROVING THE SCIENCE CURRICULUM

In relation to changing the science curriculum, the following four aspects are considered below: (a) how change is brought about; (b) how science courses should be organized across the twelve years of schooling; (c) changes in emphases on content; and (d) changes in emphases on processes.

1. *Introducing change in the science curriculum.* In introducing curriculum change in science, key issues to be considered are the definition of the intended curriculum to be covered, the allocation of class time for science, and the selection of science textbooks. In the Second IEA Science Study, information was collected concerning where decisions relating to these three issues were made within each country (Rosier and Keeves, 1991). Three general categories of response were obtained from systems: decisions made by teachers at the school or class level, and with general guidelines sometimes laid down at a regional level (T); decisions made by the school board, sometimes with general guidelines laid down at a regional level (S); and decisions made by authorities at the country or system level (C).

Table 2-4 summarizes the information recorded for 10-year-olds, 14-year-olds, and science specialist students for each of the countries that provided information with respect to the definition of the intended curriculum, the allocation of class time for science, and the selection of textbooks. The assignment of the information to one of the three categories

Table 2-4

The Locus of Decision-Making for Science Curricula, 1983-1984

Table 2-4

The Locus of Decision-Making for Science Curricula, 1983-1984

Country	Definition of Intended Curriculum in Science			Allocation of Class Time for Science			Selection of Science Textbooks		
	Pop 1	Pop 2	Pop 3	Pop 1	Pop 2	Pop 3	Pop 1	Pop 2	Pop 3
Australia	T	T	C	T	T	T	T	T	T
Canada (English)	S	S	S	S	S	S	T/S	T/S	S
Canada (French)	S	S	S	S	S	S	S	T/S	S
China	C	C	C	C	C	C	C	C	C
England	T	T	C	T	T	T	T	T	T
Finland	C	C	C	C	C	C	S	S	S
Hong Kong	C	C	C	S	T	T	T	T	T
Hungary	C	C	C	C	C	C	T/C	T/C	T
Israel	C	C	T/C	C	C	C	T	T	T
Japan	C	C	C	C	C	C	C	C	T
Korea	C	C	C	C	C	C	C	T/S	T/S
The Netherlands	-	T/C	-	-	T/C	-	-	T	-
Nigeria	C	C	C	C	C	C	C	C	C
Norway	C	C	C	T/S/C	T/S/C	C	T/S	T/S	T
Papua New Guinea	C	C	C	C	C	C	T/C	T/C	T/C
Philippines	C	C	-	C	C	-	S	S	-
Poland	C	C	C	C	C	C	C	C	C
Singapore	C	C	C	S	S	S	S	T/S	T/S
Sweden	C	C	C	T/C	T/C	C	T	T	T
Thailand	C	C	C	C	C	C	C	C	S
USA	S	S	S	S	S	S	S	S	S

T	Decision made by teachers	Pop 1	10-year-old level
S	Decision made by school board	Pop 2	14-year-old level
C	Decision made at the country or system level	Pop 3	Science specialist level

given above is indicated by the letters T (teachers), S (school based), and C (country or system authorities).

There is general consistency across the three levels for each particular country. Australia and England are the only countries for which the teachers within a school had the major say in what was taught, the time allocated, and the textbooks selected. In Australia, some state systems provided guidelines, but in England teachers had great freedom until 1989 when major changes were introduced. In Canada and the USA, school boards had the major say in determining the curriculum, time allocation, and textbooks. However, guidelines were provided in some provinces or states and only certain textbooks were authorized by certain states. In the Netherlands, Norway, and Sweden, teachers had greater freedom than in most other countries, but guidelines were laid down at the national level.

In those countries in which the teachers have the major say in determining the science curriculum and the time allocated to science, it is extremely difficult to introduce widespread change. Likewise, where school boards have the major say in such decisions, change is probably even more difficult to make. However, in centrally controlled systems, research can be conducted, public debate on change can occur, and change can be prescribed. In such systems, where the need for change is seen clearly, it can be introduced relatively quickly and effectively. There are advantages with each of the three arrangements. Nevertheless, where school boards and individual teachers have the responsibility for making the critical decisions, the introduction of widespread change inevitably must prove extremely difficult to initiate and implement. The great danger in a centrally controlled system is that the whole curriculum can be oriented in what later proves to be an unprofitable direction.

At the terminal secondary school level, university entrance examinations have a dominating influence. Even in the USA, where school boards have the major say in what is taught in schools, the examinations conducted by the nation-wide agencies have a powerful effect on science curricula and how the teaching of science is organized. As a consequence, in most systems where reform in the teaching of science is necessary and desirable, it is possible to make changes that apply across the country through the examinations conducted. However, examination bodies generally are resistant to change, particularly where there is intense competition for entry to universities and higher education.

2. *The organization of the science curriculum.* The pattern that gradually is emerging across the world is that the study of science in some form should be mandatory at all levels of schooling from Grade 1 to Grade 12. During the elementary school years, the student commonly is introduced to science as an integrated subject with content from all major fields of biology, earth science, chemistry, and physics being taught. At the lower and middle secondary school levels, some countries teach to all students an obligatory course that involves scientific content from all major fields. Differences occur between countries in the extent to which the four fields of science are taught separately or integrated in whole or in part (with biology and earth science being combined, and chemistry and physics being brought together in a physical science course). One of the factors determining which arrangement is chosen is whether the teachers have been trained and are willing to teach science as an integrated or partly integrated subject, or whether they insist on maintaining separation of the fields in their classroom teaching.

At the upper secondary school stage, specialization occurs in the subjects taken by students. Commonly, specialization involves separation into science and non-science lines, and in many countries separation occurs into academic schools and vocational schools for the final years of schooling. Generally, little science is taught formally to students in vocational schools. However, countries differ in the extent to which they offer or require students who are non-science specialists to study some science subjects. It is here that new types of science courses could have a widespread appeal and could cover the history and philosophy of science; major science-related issues; the relationships among science, technology, and society; and the methods of scientific inquiry. These courses can be highly stimulating and informative. Moreover, they provide the background knowledge and understandings that administrators and informed citizens are likely to need in the twenty-first century in a world in which the impact of science and technology on society is increasing markedly.

3. *Changes in emphases on content.* There is a constant need to prune from science courses content that is no longer central to an understanding of a field of science and, at the same time, to introduce new concepts and principles that can be shown to be of growing significance. The danger is that the curriculum will be overcrowded with excessive material so that the key ideas are not examined in relation to

each other and to their important applications. It has been argued that countries do not differ greatly in the science content which they teach to science specialist students in the fields of biology, chemistry, and physics at the terminal secondary school level. Thus, there is general consensus across countries on the topics considered to be important. However, more thought needs to be given in many countries to how there might be a concyclic or spiral development of key scientific ideas and principles across successive stages of schooling.

4. *Changes in emphases on process.* Information was presented above on the degree of emphasis given to the process dimension of the science curriculum across countries. It was argued that this aspect of the science curriculum was taught primarily through laboratory work. In addition, with the rapid expansion of scientific knowledge, there is growing recognition of the importance of problem-solving skills. As a consequence, development of the skills of scientific inquiry is likely to play an even more important role in the future than it has in the past. However, the context in which these skills are taught and the tasks employed to provide opportunities for the development of these skills might well be changed. New technology is changing the ways in which observations are examined and analyzed and, to some extent, what is observed and the way in which observations are made and measurements are recorded. Laboratory work in science must take cognizance of these changes and must be modified accordingly. The skills to be developed remain, but the approaches to their development need to be changed.

CONCLUSION

In order to plan the improvement of science courses in a changing world, it is necessary, within each particular country and each particular school, to consider the changes that have occurred over time, especially since the last major period of reform in the 1960s and 1970s. In this chapter we have placed in a worldwide context the advances that have taken place in the past, and the changing emphases that are directed toward the future. Thus, we have covered in brief not only the history of science curricula over the past 100 and more years, but we have also considered the extent to which developments have occurred more recently (for example, the Science-Technology-Society [STS] movement) in order to improve the teaching of science. The strong and sometimes passionate advocacy for reform that occurred in the 1960s led to some disillusionment and, although reform was not universal within any particular country, some significant developments took place. In the years ahead, it would seem that more gradual and systematic change is required in most countries, especially because reform is dependent on the quality of teachers, who must be trained or gradually retrained.

We have also reported some data from science studies sponsored by the International Association for the Evaluation of Educational Achievement (IEA). In particular, international findings are reported concerning participation rates, conditions of learning, and the emphasis on different content areas and process skills in various countries.

Planning to improve the teaching of science is a recurring challenge for all science educators. The advances that have emerged, which require widespread adoption to improve the teaching and learning of science, are summarized briefly below as recommendations for improving science education:

1. All students should study some science formally during all grades of schooling.
2. During the years of compulsory schooling, all students should study content from the four fields of science, biology, chemistry, earth science, and physics.
3. Greater emphasis should be placed on learning the skills of investigation and inquiry in the study of science, with the laboratory and experimentation playing an important, but not exclusive, role.
4. The relevance of science should be emphasized through greater consideration of the application of scientific principles to everyday life, technology, the production of food, and the conservation of the environment.
5. At all grade levels, relationships between science, technology, and society should be emphasized, especially for those students in the terminal grades of schooling who have concern for major science-related social issues.
6. The learning of science should be seen as a lifelong process, and not limited to the years of schooling, because scientific discovery not only is ongoing, but it is advancing at an increasing rate.

REFERENCES

Aikenhead, Glen S. "Science-Technology-Science Education Development: From Curriculum Policy to Student Learning." Paper presented at Conferencia Internacional Ensino de Ciencias Paso O Seculo XXI, Brazilia, 1990.

Aikenhead, Glen S. "What Is STS Teaching?" In *STS Education: International Perspectives on Reform*, edited by Joan Solomon and Glen Aikenhead. New York: Teachers College Press, 1994.

Armstrong, Henry E. *The Teaching of Scientific Method and Other Topics in Education*. London, England: Macmillan, 1903.

Conant, James B. *On Understanding Science*. New Haven, Conn.: Yale University Press, 1947.

Cooley, William W., and Klopfer, Leopold E. *Test on Understanding Science (TOUS)*. Princeton, N.J.: Educational Testing Service, 1961.

Eijkelhof, Harrie M. C., and Lijnse, Piet L. "The Role of Research and Development to Improve STS Education: Experiences from the PLON Project," *International Journal of Science Education* 10 (1988): 464-474.

Gee, Brian, and Clarkson, Stephen G. "The Origin of Practical Work in the English School Science Curriculum," *School Science Review* 73 (1992): 265-279.

Holton, Gerald; Rutherford, James; and Watson, Fletcher. *The Project Physics Course*. New York: Holt, Rinehart, & Winston, 1970.

Jenkins, Edgar W. "History of Science Education." In *International Encyclopedia of Education*, edited by Torsten Husén and T. Neville Postlethwaite, pp. 4453-4456. Oxford, England: Pergamon Press, 1985.

Keeves, John P., ed. *Changes in Science Education and Achievement: 1970 to 1984*. Oxford, England: Pergamon Press, 1992a.

Keeves, John P. *Learning Science in a Changing World*. The Hague, The Netherlands: International Association for the Evaluation of Educational Achievement, 1992b.

Klopfer, Leopold E. "Evaluation of Learning in Science." In *Handbook of Formative and Summative Evaluation of Student Learning*, edited by Benjamin S. Bloom, J. Thomas Hastings, and George F. Madaus, pp. 559-641. New York: McGraw-Hill, 1971.

Klopfer, Leopold E., and Cooley, William W. "The History of Science Cases for High School in the Development of Student Understanding of Science and Scientists," *Journal of Research in Science Teaching* 1 (1963): 33-47.

Mayor, Frederiko. "Address." In *Issues in Science Education: Science Competence in a Social and Ecological Context*, edited by Torsten Husén and John P. Keeves, pp. 9-17. Oxford, England: Pergamon Press, 1991.

Postlethwaite, T. Neville, and Wiley, David E. *Science Achievement in Twenty-Three Countries*. Oxford, England: Pergamon Press, 1991.

Rosenthal, Dorothy B. "Two Approaches to STS Education," *Science Education* 73 (1989): 581-589.

Rosier, Malcolm J., and Keeves, John P. *Science Education and Curricula in Twenty-Three Countries*. Oxford, England: Pergamon Press, 1991.

Skilbeck, Malcolm. *School-Based Curriculum Development*. London, England: Harper, 1984.

Solomon, Joan. "Science Education Through Science, Technology and Society." Paper presented at Conferencia Internacional Ensino de Ciencias Paso O Seculo XXI, Brazilia, 1990.

Walberg, Herbert J. "Improving School Science in Advanced and Developing Countries," *Review of Educational Research* 61 (1991): 21-69.

Welch, Wayne W. "Twenty Years of Science Curriculum Development: A Look Back." In *Review of Research in Education*, Volume 7, edited by David C. Berliner, pp. 282-306. Washington, D.C.: American Educational Research Association, 1979.

Ziman, John. *Teaching and Learning about Science and Society*. Cambridge, England: Cambridge University Press, 1980.

Chapter 3

STUDENTS' CONCEPTIONS AND CONSTRUCTIVIST TEACHING APPROACHES

Reinders Duit and David F. Treagust

. . . (T)he students had memorized everything, but they didn't know what anything meant. [Feynman, 1985, p. 192]

Science instruction, from the elementary school to the university level, is frequently disappointing as far as promoting students' understanding of science is concerned. Students are often in full command of science terminology and, for example, might be able to provide the names of animals and plants, to write down the Schroedinger equation without any difficulties, or to provide key examples when presented with formulas. However, there very often is no deep understanding behind the facade of stored factual knowledge. Understanding, as we use the term here, includes an awareness of the basic qualitative ideas in which the facts and formulas are embedded and the ability to employ that knowledge in new situations. In this context, mere retrieval of stored items from memory does not indicate understanding.

While factual knowledge about science is relatively easy to learn and present when required in examinations, it is much more difficult to achieve understanding of science. The deep-rooted reasons for these difficulties have been revealed during the past twenty years in the research under review in this chapter. The role of students' preinstructional conceptions has proven to be important in learning. Conceptions are the

Chapter consultants: Mariana Hewson (University of Wisconsin-Madison, USA) and Richard Gunstone (Monash University, Australia)

individual's idiosyncratic mental representations, whereas concepts are something firmly defined or widely accepted, in this case in science. At all ages students hold conceptions about many phenomena and concepts before these are presented in science classes. These conceptions stem from and are deeply rooted in daily experiences because they have proved to be helpful and valuable in the individual's daily life contexts.

Very often, however, these conceptions are not in accord with science concepts. Research has shown that students do not switch easily from their old preinstructional conceptions to the new science concepts taught. There are two broad reasons for this. First, students usually are quite satisfied with their own conceptions and do not see the value of the new science ideas. Second, the conceptions which are held determine substantially the subsequent learning process; students see what the teacher (or the textbook) presents through the lenses of their preinstructional conceptions and, consequently, they may not understand the science concept. Nevertheless, students sometimes can form some sort of integration of the preinstructional conceptions with newly taught science concepts, but usually their preconceptions are not altered seriously. Students often learn what is expected by rote without any understanding. How these preinstructional conceptions can be diagnosed and how teaching can be designed to take students' conceptions into account are described in more detail in Treagust, Duit, and Fraser (in press).

In the first section of this chapter we describe the constructivist view of teaching and learning science. In the second section we discuss research on students' conceptions, including specific examples from four science content areas. In the third section, sources of students' conceptions arising from sensual experiences, language, culture, and science instruction are described. In the fourth section we provide details of four innovative teaching approaches based on the constructivist perspective. In the final section we offer seven recommendations for improving science education.

THE CONSTRUCTIVIST VIEW OF TEACHING AND LEARNING SCIENCE

History of Research on Students' Conceptions

Although awareness that students' preinstructional conceptions play a key role in the learning process is not a new insight, it has been given serious attention in science education research only in the past twenty years. Studies of students' conceptions in science were carried out as early as the beginning of this century (for example, Hall and Browne, 1903), and in the middle of the century Oakes (1947) presented an extensive review of studies in this field. But it was only in the middle of the 1970s that this issue was given the priority in science education research that is needed for the improvement of science instruction. It is not by accident that the research and development that started then still flourishes today. On the one hand, the limited success of the science curricula that were designed and evaluated with huge efforts in the 1960s and early 1970s forced science educators to rethink science instruction. On the other hand, the view that the acquisition of new knowledge is very much influenced by conceptions already held became a key idea in a variety of fields. In the domain of philosophy of science, for instance, the idea that conceptions guide observation and determine understanding became prominent. Hanson's (1965) idea of theory-laden observation, in which observations are shaped considerably by the conceptions which an individual holds, and Kuhn's (1970) seminal analysis of the impact of old ideas on the development of new ones in the history of science, can be taken as paradigmatic examples.

Within cognitive science and information-processing theories, there is the view that the conceptions held by each individual guide understanding. It appears also that developments in other fields (such as research on self-organizing systems) contributed to the appeal of this view in a large variety of domains. In science education, as well as in many other fields, this idea is labeled the "constructivist view" and it has been a significant influence in assisting our understanding of students' learning difficulties and in developing new teaching and learning approaches.

Basic Principles of the Constructivist View

At the heart of constructivism is the idea that knowledge about the world outside is viewed as human construction. A reality outside the individual is not denied; rather, it is claimed only that all we know about reality is our tentative construction. Accordingly, learning is not viewed as transfer of nuggets of truth, as Kelly (1955) put it, but as active construction. It is the learner who constructs, or even creates, his or her knowledge on the basis of the knowledge already held. In addition, there are social aspects of the construction process; although individuals have to construct their own meaning of a new idea, the process of constructing meaning always is embedded in a particular social setting of which the individual is part.

If a communication situation between students and a teacher is considered from the constructivist viewpoint, it is apparent that there is a circle of understanding that is well known from pedagogy. If the teacher asks a question and students try to understand it, they are able to do this only from their perspective and on the basis of the conceptions that they hold. If these conceptions are different from those of the teacher, and this usually is the case, the students make sense of the question in a way different from the teacher's way; the answer the students might give is interpreted by the teacher from his or her point of view. An endless circle of misunderstanding can occur in such communication situations, and these incidents frequently occur in teaching and learning. Neither the teacher nor the student can be sure that he or she really understands the other. What a teacher calls a student's wrong answer actually can be his or her construction on the basis of his or her conceptions. Therefore the teacher must be aware that students usually argue from a vantage point that is different from the teacher's. This issue is significant for research on students' conceptions because what researchers call students' conceptions are actually the researcher's conceptions of the students' conceptions.

Conceptual Change and Conceptual Growth

Distinctions have been made between two complementary kinds of learning, *conceptual growth* and *conceptual change*. Other authors refer to *conceptual addition* and *conceptual capture*. Conceptual growth refers to enlargements of the conceptual network in such a way that one's previous knowledge and its connections, for the most part, remain intact. The term conceptual change often is used loosely and with slightly different meanings, but mostly it refers to learning in which major parts of the existing conceptual network are reorganized. Research on students' conceptions has shown that learning of the conceptual change kind is required for enhancing science knowledge. A model of conceptual change espoused by Hewson and Hewson (1992) proposes first that there are conditions that need to be met in order for a person to experience conceptual change, and second that the person's conceptual ecology provides the context in which the conceptual change occurs, influences the change, and gives it meaning. Conceptual ecology comprises a person's already existing conceptions and beliefs (including epistemological and metaphysical beliefs) and the multiplicity of interrelations between them.

In discussing the meaning of conceptual change, Hewson and Hewson (1992) claim that change does not mean erasing or extinguishing students' conceptions totally, for the following reasons. First, students' conceptions have been found to be valuable in most everyday contexts and are perceived to be valuable in future specific situations. Second, most lay adults, as well as experts, hold at least major features of these conceptions in fields that are not familiar to them. Finally, research has shown that it is impossible to extinguish old conceptions totally and, therefore, a coexistence view appears to be much more appropriate. The aim of science instruction from this perspective is not to erase students' conceptions, but to work out the contexts in which these conceptions are limited and in which the science conceptions are more valuable.

The Role of Experiments in Learning Science

Experiments play a key role in teaching and learning science in traditional and constructivist settings. However, within the constructivist perspective the role of experiments in the learning process is viewed with more caution than in traditional approaches. The constructivist view holds that there is no objective observation and that every observation is theory-laden, that is, observations are determined by the conceptions held (Hanson, 1965). Indeed, it is an everyday experience that different people who report the same event actually "observe" different features of this event or even "observe" the same feature differently. Empirical studies in the field of psychology indicate that humans in general tend to observe only what fits their conceptions and to ignore counterexamples. Results of studies in science education show clearly that students often do not see what is obvious from the point of view of the presenter of the experiment.

Cognitive Conflict

In order to guide students from their conceptions to the science concept, cognitive conflict situations can be arranged to contradict students' predictions. Nevertheless, even if students observe the features which the teacher intends, one single piece of empirical evidence usually does not shake students' conviction that their view is right; consequently, they try to eliminate the counterevidence by all sorts of arguments. For instance, students claim that the specific situation of the experiment is responsible for the unexpected outcome. In a study by Tiberghien (1980), a student had to find out whether an ice block covered with wool melts more quickly than an ice block covered with aluminum foil. The student was of the opinion that the ice block covered with wool would melt first, because wool keeps a person warm and hence wool gives warmth. The existence of this conception is well known from other studies. The empirical evidence that the ice block covered with aluminum foil melts first did not shake this conviction. The behavior of Tiberghien's student, and of other students in other studies, is reminiscent of the behavior of scientists as reflected in the history of science. One single counterexample usually does not lead to giving up a previously held theory or idea (Kuhn, 1970).

Beyond Content Issues

Research into students' preinstructional conceptions in the middle of the 1970s initially involved students' conceptions of content, especially related to processes such as seeing or the phenomenon of photosynthesis. In the early 1990s, most emphasis still is given to this type of conception, which continues to be of key interest in science instruction. However, the constructivist view led to the insight that many other conceptions held by students and teachers also influence the learning process. The most important conceptions are outlined below and in more detail in Duit (1991).

Meta-knowledge. Meta-knowledge involves conceptions of the nature of content knowledge. How students conceive the nature of knowledge is an important yet often neglected issue. Learning difficulties revealed in empirical studies appear to be due partly to misunderstandings of the status of students' conceptions. Studies indicate that most students and many teachers are naive realists who view scientific knowledge as a faithful copy of the world and not as a tentative human construction. Hence, there is a common misunderstanding that scientific investigations are based mainly on objective observation of nature and not on the construction of explanations about nature.

Conceptions about the aims of instruction and the purpose of particular teaching events. Teachers' and students' ideas about the aims of instruction very often are not in accord. Whereas most teachers have a long-term perspective which involves placing a single event within a structured sequence of related events, many students appear to lack such a long-term perspective. For instance, students can view an experiment as a single event unrelated to others, and hence lack an appropriate framework that could guide their investigations.

Conceptions about the learning process. This is a key issue from the constructivist perspective because the conceptions of both teachers and students determine the teaching and learning processes. Empirical studies strongly support that the view of learning in which students are considered as "empty vessels" to be filled with knowledge is still dominant among both teachers and students. Attempts to change teachers' and students' views of the learning process and to provide them with efficient learning tools, or "metacognition" (Gunstone, 1992), have been successful in providing powerful teaching and learning tools as well as promoting students' understanding in science.

The role of the teacher. The constructivist view of learning can facilitate understanding of science content in a much more appropriate way than traditional approaches. But, in addition to a focus on science instruction, a constructivist view has important consequences for the role of the teacher in a classroom. The teacher is viewed as a facilitator of knowledge construction (that is, as a guide in students' individual construction processes) rather than as a person who transfers knowledge to the brains of the students. Teachers and students are seen as partners in the teaching and learning situation. Consequently, students are given more command of their own learning and more responsibility for it. Relations between students and teachers are more symmetrical than in teacher-dominated classrooms. The classroom climate of a constructivist classroom accordingly can be more satisfying for many teachers and many students (Taylor and Fraser, 1991). The climate in constructivist classrooms appears to be appealing relative to other approaches because science instruction is more meaningful. This is consistent with Science-Technology-Society (STS) and gender-inclusive approaches addressed in chapters 1 and 9, respectively, in the present volume and with Fensham's (1986) description of "science for all," which explicitly includes constructivist ideas, STS, and gender.

RESEARCH ON STUDENTS' CONCEPTIONS

Most research still emphasizes students' conceptions about content as is evidenced by the bibliographies of Pfundt and Duit (1991) and of Carmichael, Driver, Holding, Phillips, Twigger, and Watts (1990). Pfundt and Duit's bibliography, which is updated continually and in which there are currently around 1,200 entries of this kind, reveals that about 70 percent of the studies have been carried out in the field of physics as shown in table 3-1. Although the number of studies in the physics domain is quite substantial, there are still major topics that await further investigation, including elementary physics, sound and magnetism, and topics of modern physics. In biology and chemistry, there is a need for further studies on students' conceptions in many areas.

Terminology

In addition to the term *students' conceptions*, there are many other terms used depending on the authors' views of the nature of knowledge. The term *alternative frameworks* suggests that students' conceptions are valuable in many everyday contexts, and are not just wrong and hindrances to learning. Other terms which indicate that students' conceptions should be taken seriously are *children's science* or *children's mini-theories* (suggesting that children design theories in a way that is similar to scientists). The term *misconception* refers to a conception that is wrong from the science point of view. Although use of this term still can be found, it now tends to be employed to indicate wrong conceptions (from the scientific point of view) that have been caused by science instruction itself.

Examples of Students' Conceptions

Findings from research on students' conceptions in selected content areas are summarized briefly below. Research generally has shown that students' conceptions are surprisingly similar in different student age groups, that the same conceptions are held by lay adults, and that some

Table 3-1

Number of Studies of Students' Conceptions in Different Science Content Areas

General Area	Number of Studies	Specific Topics
Mechanics	281	Force and motion; work, power, energy; speed, acceleration; gravity; pressure; density; floating, sinking
Electricity	146	Simple, branched circuits; topological and geometrical structure; models of current flow; current, voltage, resistance; electrostatics; electromagnetism; danger of electricity
Heat	68	Heat and temperature; heat transfer; expansion by heating; change of state, boiling, freezing; explanation of heat phenomena in the particle model
Optics	69	Light; light propagation; vision; color
Particles	60	Structure of matter; explanation of phenomena (e.g., heat, states of matter); conceptions of the atom; radioactivity
Energy	69	Energy transformation; conservation; degradation
Astronomy	36	Shape of the earth; characteristics of gravitational attraction; satellites
"Modern" Physics	11	Quantum physics; special relativity
Chemistry	132	Combustion, oxidation; chemical reactions; transformation of substances; chemical equilibrium; symbols, formulas; mole concept
Biology	208	Plant nutrition; photosynthesis; osmosis; life; origin of life; evolution; human circulatory system; genetics; health; growth

Based on Pfundt and Duit (1991)

teachers hold the same or similar conceptions, that is, conceptions not in accord with science concepts. (See the bibliography of Pfundt and Duit, 1991.) As previously mentioned, traditional instruction often is able to change conceptions only superficially. In the following sections, therefore, a description is given of major features of students' conceptions that persist after traditional science instruction.

Conceptions of light. There is a substantial number of studies of students' conceptions of light (see table 3-1). From the many findings available, only students' conceptions of the process of seeing are considered here. The physicist explains how one sees a light source (such as a candle) or a lit object (such as a tree) in the same way, that is, the surfaces of light sources emit light or lit objects reflect light. The light travels in a straight line and it passes from the object to the eyes of the person; an image is formed on the retina that is interpreted by the brain. However, students' conceptions have been found to be significantly different from this scientists' view (Jung, 1989). Within students' alternative frameworks, there is no idea of light travelling in the same way as in physics. Second, there are two kinds of light conceptualized by students. Bright objects, referred to as light sources in physics, can be seen if the observer turns her or his eyes toward them. Objects that provide this kind of light can make other objects bright. The second type of object can be seen under the same conditions as the light-providing ones, but these objects themselves are not able to make other objects bright. The other kind of light is a feature of objects that can be called their *brightness*; it illuminates the objects but stays there.

These students' conceptions of light are obviously in stark contrast to physics concepts. For instance, the idea that light travels is not held by most students. But this is not surprising because there simply is no evidence from everyday experiences with light phenomena to support this idea. While science instruction often attempts to guide students to the science view via experimental observation, it has to be admitted that the idea of light travelling is chiefly a matter of theory and thinking rather than a matter of observation, at least on the basis of experiments to which students have access.

Conceptions of the particulate nature of matter. The particle model plays an important role in science teaching, although it is a difficult concept because a model of a micro-world is constructed that is not accessible to our senses. To facilitate understanding of this model, analogies with the macro-world usually are drawn (for example, one can think of the particles in terms of small spheres). Such analogies to the macro-world (that is, to the world familiar to the student) are often like a Trojan Horse because students have difficulties in understanding that the micro-world of the particles is totally different from the macro-world. (See the summary of studies on the particle model by Andersson, 1990.) Accordingly, students tend to attach features of objects in the macro-world also to the particles. In many students' views, sulfur particles are yellow because sulfur is yellow. Also, hot particles expand when heated and, if they rub against one another, heat is produced. There are many further examples of this kind. Students often do not understand that particles in the micro-world never stop moving because bodies in the macro-world come to rest after a while because of friction. That there is no friction in the micro-world is a difficult idea for students who are told that the particles behave in a way somewhat similar to bodies in the macro-world. Indeed, a very similar conclusion concerning the significance of empirical evidence holds here as in the case of students' ideas of light. Again, the particle model does not "spring out" of appropriately arranged observations but, rather, the model is a matter of theory and thinking.

Conceptions of energy. Energy is among the concept labels that are used in science differently than in the everyday world. The word "energy" has been used frequently in daily life contexts only since energy problems have been discussed seriously by the public. The use of the word "energy" in daily life (for example, in discussions of energy supply issues) is mirrored by students' conceptions of energy. Energy frequently is seen as a universal kind of fuel which can be produced from certain sources like oil, gasoline, coal, sun, wind, or flowing water (Duit and Häussler, 1994). Accordingly, for many students, energy is an industrial product which can be obtained from certain raw materials needed in order to operate machines. Energy also is seen as a luxury item; for example, students tend to see energy not as a basic necessity of all processes, but only of those technical processes that make our lives more comfortable. Students' conceptions of energy are influenced strongly by everyday language and there have been some differences reported between languages. In the English language, energy frequently is affiliated with living beings, especially with humans and also with food, but these linkages are weaker in the German language. Viewed from the energy concept in science, the idea of energy conservation is missing in students' conceptions. However, the word energy as used in daily life is quite close to the science concept of free energy that includes ideas of energy degradation.

Conceptions of plant nutrition. Research into students' conceptions in biology has been much less common than in physics, although the biological concepts involved in plant nutrition have been the focus of a number of studies during the past decade. This is not surprising because concepts about plant nutrition are central to biology education and are included in most secondary school curricula. A common outcome of these studies is that students appear to have limited understanding of the relationship between photosynthesis and other physical and chemical processes, such as water uptake and respiration, carried out by plants. Indeed, even students in the last two years of secondary school lack a clear understanding of the relationship between photosynthesis and respiration and the role of energy in the maintenance of plant metabolism. Generally, students believe that plants obtain their food from the environment, rather than by manufacturing it internally, and that food for plants is anything that is taken in from the outside, such as water, minerals, and air. The expression of these ideas appears to be influenced by prior everyday observations of plants needing water, sunshine, and fertilizers. However, an examination of the way in which plant nutrition is taught indicates that these prior ideas are not used by teachers in promoting students' understanding of photosynthesis as a carbohydrate-producing process and of how their own ideas can be used to build up scientific knowledge. Indeed, research suggests that students' knowledge can be enhanced greatly when their own ideas are incorporated actively into classroom activities.

SOURCES OF STUDENTS' CONCEPTIONS

We turn now to sources of students' conceptions about science content (that is, those labeled as alternative frameworks or misconceptions) which can arise as a result of sensual experiences, language experiences, cultural background, peer groups, and the mass media, as well as from science instruction.

Sensual experiences. Students' sense impressions from daily life experiences are a major source of their conceptions with regard to phenomena like heat, light, forces, motion, combustion, rusting or other chemical changes, and the growth of plants and animals. These conceptions, constructed in an attempt to make sense of the phenomena, appear to be the ones that are most resistant to change through instruction. They are rooted deeply in personal experiences, merged with feelings, and usually are successful in providing explanations for everyday situations.

Language experiences. Everyday language experiences significantly influence conceptions of natural phenomena. First, our language provides views of some phenomena that have been outdated in science for a long time. The sentence "the sun rises" is a significant example because it ignores in some way the view that the world revolves, a view initiated by Copernicus, Kepler, Newton, and others. "The sun rises" conserves the ancient view of the sun (the cart of helios) traveling in celestial spheres around the earth. Second, many terms for science concepts, such as "force" and "energy," are in use also as words in everyday language, usually with different meanings from science. "Force" in everyday language has multiple meanings, including the meaning of a power possessed by a living being or a machine. This provides considerable contrast to the physics meaning in which force is a measure of the strength of interaction between bodies.

Whereas mutual sensual experiences in different cultural contexts appear to be responsible for similarities of some conceptions across different cultures, language appears to be responsible for some differences that research has revealed. These differences mainly appear to be caused by different conceptualizations of natural phenomena, like heat, as provided by different languages.

Cultural background, peer groups, and mass media. Learning about the issues taught in science also takes place often in everyday contexts outside science classrooms. Major basic influences, namely, sensual and language experiences, already have been mentioned. In daily communications with friends, parents, and others, as well as from reading and the mass media, students hear, see, and use aspects of both science knowledge and alternative conceptions.

Science instruction. There is substantial research evidence indicating that science instruction either supports students' old (alternative) conceptions or even causes new misconceptions. There are several reasons for these pitfalls. First, misconceptions are provided by teachers and textbooks and other teaching media. Teachers also can hold major misconceptions, especially teachers who do not have adequate background and training in science. Further, significant errors can be found in textbooks and other media. Some of these errors appear to be rooted in a long-standing tradition of copying ideas from one book to another without critical proof. For example, in a leading German science museum, there was for many years a display that presented a false conception of the action and reaction ideas that are central to the Newtonian concept of force: the acting force (action) and the reacting force (reaction) in the display were acting on the same body whereas, in the Newtonian view, the forces act on different bodies. Science instruction can also cause misconceptions when information that is basically correct (from the science point of view as given by teachers or textbooks) is misinterpreted by students. Students do this when they attempt to make sense of the provided information from their point of view, which often is considerably different from the science point of view.

INNOVATIVE TEACHING APPROACHES

This section briefly describes some innovative teaching approaches that are based on the constructivist perspective (see the review of Scott, Asoko, and Driver, 1992, and the book by Treagust, Duit, and Fraser, in press). A recent meta-analysis of seventy studies of this kind by Guzetti and Glass (1992) suggested that teaching approaches that offend or challenge students' preinstructional conceptions generally are significantly superior to approaches that do not take students' conceptions into account. Indeed, the success of these new approaches generally is very promising.

Starting from students' point of view. An old pedagogical principle involves starting instruction from the student's point of view. But only now are there opportunities to set this principle into practice coherently because common student conceptions are known in numerous main areas of science and this knowledge can be used by curriculum developers and teachers. Also, powerful new methods of assessment, which allow teachers to investigate their students' conceptions and understanding, are now available (White and Gunstone, 1992).

Changing from students' conceptions to science concepts can be accomplished by two types of teaching, namely, continuous or discontinuous approaches. These can be seen as analogous to Kuhn's (1970) evolutionary and revolutionary changes in the conduct of science. Continuous approaches start with aspects of students' conceptions that already are in general accord with science concepts or that can be reinterpreted from the science point of view. Discontinuous approaches usually contain, at some stage, the cognitive conflict strategies which are reviewed by Scott, Asoko, and Driver (1992). There are three primary kinds of cognitive conflict: (a) between students' predictions in an experiment and its actual outcome; (b) between students' conceptions and the teacher's conceptions; and (c) between the conceptions of different students. Although cognitive conflict strategies have been found to be effective, they have to be used with some caution. The crucial point is whether or not students really see the conflict: what might appear as a conflict in the teacher's opinion might not be seen as a conflict by students from their perspective.

Constructivist teaching sequence. The teaching sequence described by Driver (1988), as shown in figure 3-1, is paradigmatic for many other approaches. Through some kind of elicitation of students' conceptions (involving exploring their own ideas, discussing the differences among ideas of different students, carrying out experiments, and trying to explain the observed phenomena), students usually become aware of their own and others' points of view. During the restructuring phase, students' ideas can be clarified, challenged, and exchanged through discussions with others, or the teacher can promote conceptual conflict through the use of a disconfirming experiment or demonstration. The scientific view can be introduced by the students or the teacher and the different ideas evaluated against experience, through experiment, or by thinking through the implications. In the application phase, students are given the opportunity to consolidate and reinforce new conceptions by using them in both familiar and novel situations. In the review phase, students compare their new views with earlier ones.

A key feature of this constructivist teaching sequence is a phase involving the contrasting of students' ideas and the science conception. As previously mentioned, the teacher is seen as a facilitator of students' construction processes and not as a transmitter of the science view. However, there are certain dangers in this phase. For example, students could be unable to see the differences between their view and the new

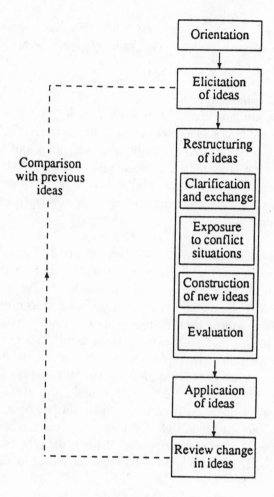

Figure 3-1

**An Example for a Constructivist Teaching Sequence
(from Driver, 1988)**

one, and some students, especially younger ones, would prefer to know the right answer than to play around with ideas.

Conceptual change model. This model, which has achieved a great deal of attention in recent years, as illustrated by the review of Hewson and Thorley (1989), emphasizes four conditions for conceptual change to take place: there has to be *dissatisfaction* with existing ideas; and the new conception must be *intelligible*, initially *plausible*, and *fruitful*. The first and the last conditions appear to be the most difficult ones to address adequately because students frequently are satisfied with their everyday conceptions and there is often no dissatisfaction with the old ideas from the students' point of view. Further, it is not easy to persuade students that the new science conceptions are more fruitful than the old ones. The three conditions—intelligible, plausible, and fruitful—are indicators of what Hewson and Hewson (1992) call the "status" of a conception. From the perspective of this model, it is the aim of science instruction to increase the status of science conceptions and not to diminish them.

Bridging analogies. Brown and Clement (1989) have developed an approach that tries to find a continuous passage from the students' conceptions to the science concept. This approach starts with those aspects of students' existing conceptions that are mainly in accord with the science view. The authors then employ a series of intermediate situations as stepping stones that are designed as bridging analogies. Figure 3-2 presents a paradigmatic example. The situation to be explained is that of a book lying on a table. Many students have difficulties in accepting, not only that the book exerts a force on the table, but also that the table exerts a force on the book (that is, the table pushes back, so to speak). Brown and Clement argue that the analogous situation at the left hand side of Figure 3-2 triggers the correct idea, namely, that the spring pushes back on the book when the book exerts a force on the spring.

Metacognitive approaches. As outlined in the first part of this chapter, students' and teachers' conceptions of the learning process play a key role in learning from the constructivist perspective. A number of approaches address this issue. Novak and Gowin (1984) propose the use of *concept maps* for this purpose and these maps are now used widely as methods to probe students' understanding of science concepts (White and Gunstone, 1992). The basic idea of the maps is that students write down the key concepts of an area and indicate the ways in which the concepts are interrelated. Novak and Gowin claim that by using concept maps students become aware of their understanding and hence increase their meta-cognitive skills in general. It also has been shown that, after

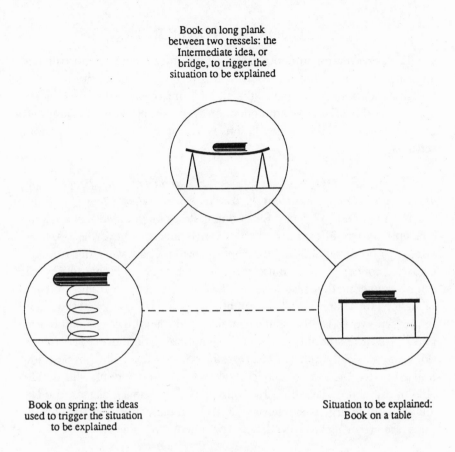

Book on long plank
between two tressels: the
Intermediate idea, or
bridge, to trigger the
situation to be explained

Book on spring: the ideas
used to trigger the situation
to be explained

Situation to be explained:
Book on a table

Figure 3-2

**An Example of a Bridging Analogy
(adapted from Brown and Clement, 1989, p. 240)**

some training, twelve-year-old students are able to use the language and ideas of the conditions of conceptual change (namely, intelligible, plausible, and fruitful) to reflect upon their learning processes while dealing with science content. Another approach for improving students' and teachers' views of learning and for improving their meta-cognitive skills in science instruction is presented by Baird and Mitchell (1986). Their approach involves, for example, asking students to keep a "learning diary" and to regularly complete questionnaires that lead to reflection about their learning progress following lessons.

RECOMMENDATIONS

The research on students' conceptions and innovative constructivist approaches to teaching reviewed in this chapter suggests the following recommendations to policymakers and administrators for improving science education. Each recommendation is presented with an example to illustrate how the recommendation can be implemented in classroom settings.

1. *Science teaching should start with the student's everyday conceptions and point of view, rather than the teacher's or scientist's view.*

For the energy concept, for example, it has been noted that students' conceptions are different from the science concept. Nevertheless, there appear to be resemblances that could provide a pathway from students' everyday conceptions, dominated by the idea of degradation, toward the science concept that also includes the idea of conservation. Duit and Häussler (1994) argue that the traditional path via "energy as the ability to perform work" is totally unacceptable, and they provide an alternative that starts from students' everyday conceptions. These everyday conceptions stay basically intact, but are reinterpreted from the physics view. Subsequently, the energy concept achieved by the students still is valuable in explaining daily life situations (that is, in science classes, the language about energy is similar to that used in daily life), and the conceptions also are in accord with the energy concept in science.

2. *New curricula should be developed to consider students' conceptions, present science in a social context, and address gender issues in order to promote students' understanding, especially in difficult fields for which students' conceptions are in sharp contrast to the science concept.*

An approach that is likely to lead to more meaningful science education is a merging of insights from the constructivist field with those from the fields of STS and gender, reviewed in chapters 2 and 9, respectively, in this volume. While new science curricula usually are introduced by means of a new textbook, until recently the textbook has been given little attention within constructivist research because this medium usually is viewed as not being well suited to enhancing students' understanding. However, a major part of science learning undoubtedly will remain dependent on some sort of textbook, at least in higher secondary grades.

There is useful research on reading science texts indicating that text-books can be improved considerably from a constructivist perspective (Guzetti and Glass, 1992). These findings suggest that a textbook that challenges students' conceptions in some way can lead to better under-standing compared to one that does not provide such challenges.

3. *Assessment procedures should be changed from an emphasis on recall of factual knowledge to a focus on understanding.*

Alternative assessment procedures, such as performance tasks and portfolio-like processes, are needed for the assessment of thinking rather than the possession of information (Wolf, Bixby, Glenn, and Gardner, 1991). Assessment by performance tasks requires students to write, read, and solve problems in genuine rather than artificial ways, whereas portfolio-based assessment involves structured sampling of a student's work from, for example, a whole year's problems in different areas of science content. Assessment also should include qualitative questions about scientific phenomena, as illustrated by Treagust's (1988) two-tier multiple-choice items in which the first tier of choices examines factual knowledge and the second tier of choices examines the reasoning behind those choices.

4. *Because different students have different experiences and knowledge bases, new media, such as computer software, need to be developed so that students can be provided with more opportunities to relate and construct their own knowledge in ways that are appropriate for them.*

There have been several attempts to develop new media that could help students to construct science concepts. Compared with more traditional attempts, these approaches do not try to improve learning just by providing one new teaching aid. Instead, these new media usually are integrated into a comprehensive constructivist approach that includes developments with both computers and textbooks.

5. *Because both science teachers' and their students' understanding of science can be enhanced by an awareness that science is a socially constructed activity, meta-knowledge of science should be given more emphasis in the science curriculum.*

For example, ideas from the philosophy of science and the history of science can be used to improve students' meta-knowledge of science

(Matthews, 1988). A growing literature in journals on science teaching provides ideas and case studies that science teachers can use to inform themselves as well as to see how such ideas could be used in their teaching.

6. *Preservice and practicing science teachers should be introduced to constructivist ideas of teaching and learning so that they become aware that the teacher's role is not to transmit knowledge but to facilitate student learning.*

Some new models of teacher education incorporating constructivist ideas have enhanced both the teachers' view of the learning process and teachers' philosophy of science ideas (see Tobin's chapter in this volume). In addition, science teachers need to experience learning science in higher education institutions in such a way that understanding science, and not the ability to recall factual knowledge, is the focus. This approach is particularly important because teachers tend to teach in the way they were taught themselves and not necessarily in the way in which they have been told to teach.

7. *Relevant research results about student conceptions should be communicated to teachers, curriculum developers, assessment writers, and teacher educators to inform and guide improvement in their practice.*

The number of existing summaries about students' conceptions in science and constructivist teaching approaches written specifically for teachers and curriculum developers is still very small. One approach to making research findings more accessible to science and mathematics teachers throughout Australia is the production of a series of eight-page publications, entitled *What Research Says to the Science and Mathematics Teacher* (Fraser, 1990), which are sent to schools nationally at regular intervals.

SUMMARY

Research on students' conceptions has revealed that the outcomes of science instruction often are limited to recall of factual knowledge; understanding of science frequently is missing. Consequently, students usually retain their everyday conceptions and do not gain faith in the value of science concepts. We have reviewed constructivist teaching and learning, research on students' conceptions with specific science examples, sources contributing to students' conceptions, and innovative approaches by which these conceptions can be challenged. Our purpose has been to provide administrators and decision makers with an appreciation of developments in this important field of research. Further support should be given to this line of research in order to achieve the progress that appears to be possible. In particular, there is a lack of the replication studies which are common in other areas of scientific research.

REFERENCES

Andersson, Bjorn. "Pupils' Conceptions of Matter and its Transformation (age 12-16)," *Studies in Science Education* 18 (1990): 53-85.

Baird, John R., and Mitchell, Ian J. *Improving the Quality of Teaching and Learning—An Australian Case Study*. Melbourne, Victoria: Monash University, 1986.

Brown, David E., and Clement, John. "Overcoming Misconceptions Via Analogical Reasoning: Abstract Transfer Versus Explanatory Model Construction," *Instructional Science* 18 (1989): 237-261.

Carmichael, Patrick; Driver, Rosalind; Holding, Brian; Phillips, Isobel; Twigger, Darryl; and Watts, Michael. *Research on Students' Conceptions in Science—A Bibliography*. Leeds, England: University of Leeds, 1990.

Driver, Rosalind. "Theory into Practice II: A Constructivist Approach to Curriculum Development." In *Development and Dilemmas in Science Education*, edited by Peter J. Fensham, pp. 133-149. London: Falmer Press, 1988.

Duit, Reinders. "Students' Conceptual Frameworks: Consequences for Learning Science." In *The Psychology of Learning Science*, edited by Shawn Glynn, Russell Yeany, and Bruce Britton, pp. 65-85. Hillsdale, N.J.: Erlbaum, 1991.

Duit, Reinders, and Häussler, Peter. "Learning and Teaching Energy." In *The Content of Science: A Constructivist Approach to Its Teaching and Learning*, edited by Peter Fensham, Richard Gunstone, and Richard White, pp. 185-200. London, England: Falmer Press, 1994.

Fensham, Peter J. "Science for All," *Educational Leadership* 44 (1986): 18-23.

Feynman, Richard P. *Surely, You're Joking Mr. Feynman*. New York: Bantam Books, 1985.

Fraser, Barry J. "Professional Development Activities of the National Key Centre for School Science and Mathematics," *South Pacific Journal of Teacher Education* 18, no. 1 (1990): 65-74.

Gunstone, Richard F. "Constructivism and Metacognition: Theoretical Issues and Classroom Studies." In *Research in Physics Learning: Theoretical Issues and Empirical Studies*, edited by Reinders Duit, Fred Goldberg, and Hans Niedderer, pp. 129-140. Kiel, Germany: Institute for Science Education, University of Kiel, 1992.

Guzetti, Barbara, and Glass, Gene. "Promoting Conceptual Change in Science: A Comparative Meta-Analysis of Instructional Interventions from Reading Education and Science Education." Paper presented at the Annual Meeting of the American Educational Research Association, San Francisco, 1992.

Hall, G. Stanley, and Browne, C. E. "Children's Ideas of Fire, Heat, Frost and Cold," *Pedagogic Seminar* 10 (1903): 27-85.

Hanson, Norwood R. *Patterns of Discovery*. Cambridge, England: Cambridge University Press, 1965.

Hewson, Peter W., and Hewson, Mariana G. A'B. "The Status of Students' Conceptions." In *Research in Physics Learning: Theoretical Issues and Empirical Studies*, edited by Reinders Duit, Fred Goldberg, and Hans Niedderer, pp. 59-73. Kiel, Germany: Institute for Science Education, University of Kiel, 1992.

Hewson, Peter W., and Thorley, N. Richard. "The Conditions of Conceptual Change in the Classroom," *International Journal of Science Education* 11 (1989): 541-553.

Jung, Walter. "Phaenomenologisches vs. Physikalisches Optisches Schema als Interpretationsinstrumente bei Interviews" [Phenomenological Versus Physical Optical Patterns as Interpretation Instruments in Interviews], *Physica Didactica* 16 (1989): 35-46.

Kelly, George A. *The Psychology of Personal Constructs*, vols. 1, 2. New York: W. W. Norton, 1955.

Kuhn, Thomas S. *The Structure of Scientific Revolutions*. Chicago: University of Chicago Press, 1970.

Matthews, Michael. "A Role for History and Philosophy in Science Teaching," *Educational Philosophy and Theory* 20 (1988): 67-81.

Novak, Joseph, and Gowin, D. Bob. *Learning How to Learn*. Cambridge, England: Cambridge University Press, 1984.

Oakes, Mervin E. *Children's Explanations of Natural Phenomena*. New York: Teachers College Press, 1947.

Pfundt, Helga, and Duit, Reinders. *Bibliography—Students' Alternative Frameworks and Science Education*. Kiel, Germany: Institute for Science Education, University of Kiel, 1991.

Scott, Philip H.; Asoko, H. M.; and Driver, Rosalind H. "Teaching for Conceptual Change: A Review of Strategies." In *Research in Physics Learning: Theoretical Issues and Empirical Studies*, edited by Reinders Duit, Fred Goldberg, and Hans Niedderer, pp. 310-329. Kiel, Germany: Institute for Science Education, University of Kiel, 1992.

Taylor, Peter C., and Fraser, Barry J. "CLES: An Instrument for Assessing Constructivist Learning Environments." Paper presented at the Annual Meeting of the National Association for Research in Science Teaching (NARST), Fontana, Wis.: 1991.

Tiberghien, Andree. "Modes and Conditions of Learning—An Example: The Learning of Some Aspects of the Concept of Heat." In *Cognitive Development Research in Science and Mathematics: Proceedings of an International Seminar*, edited by W. Fred Archenhold, Rosalind Driver, Antony Orton, and Colin Wood-Robinson, pp. 288-309. Leeds, England: University of Leeds, 1980.

Treagust, David F. "The Development and Use of Diagnostic Instruments to Evaluate Students' Misconceptions in Science," *International Journal of Science Education* 10 (1988): 159-169.

Treagust, David F.; Duit, Reinders; and Fraser, Barry J., eds. *Teaching and Learning in Science and Mathematics*. New York: Teachers College Press, in press.

White, Richard T., and Gunstone, Richard F. *Probing Understanding*. London, England: Falmer Press, 1992.

Wolf, Dennie; Bixby, Janet; Glenn, John, III; and Gardner, Howard. "To Use Their Minds Well: Investigating New Forms of Student Assessment." In *Review of Research in Education*, edited by Gerald Grant, pp. 31-74. Washington, D.C.: American Educational Research Association, 1991.

Chapter 4

INSTRUCTIONAL STRATEGIES

Avi Hofstein and Herbert J. Walberg

In the teaching of science, curriculum materials and instructional strategies ideally should be tailored to the aptitudes of the students. The traditional objective is to create an effective classroom in which (a) students are given opportunities to interact physically with instructional materials whenever possible, through handling, operating, and practicing; (b) effort is made by the teacher to provide materials and instruction that give reality and concreteness to scientific concepts; (c) and teachers vary instructional strategies, materials, and classroom procedures with the aim of increasing the effectiveness of their teaching.

The term "instructional strategy" refers to the way in which a science teacher uses materials, media, setting, and behaviors to create a learning environment that fosters desirable outcomes. Instructional strategies can be located on a continuum, one end of which is *teacher-centered* (that is, with the teacher being active and the student being less active, but not necessarily intellectually passive), and the other end being *student-centered*. Strategies that are student-centered include laboratory activities, inquiry techniques, small-group discussion, individualized learning, computer simulations, and field trips. Strategies that are teacher-centered include lectures, classroom discussions, demonstrations, and questioning techniques.

Increased student activity does not necessarily imply reduced teacher activity. A lecture, for example, could involve either factual questions or deep questions and ample response time. The amount of

Chapter consultants: Richard Kempa (University of Keele, England) and Geoffrey Giddings (Curtin University of Technology, Australia).

teacher directness also is relevant. With a direct method, science is communicated by the teacher to the students, whereas with an indirect method the teacher plays the role of a facilitator, guide, or catalyst in drawing upon student experiences and insights. The use of different strategies and methods clearly requires shifting the roles and responsibilities of both students and teachers (Holdzkom and Lutz, 1989).

The purposes in this chapter are to consider a variety of student-centered instructional strategies such as inquiry learning, computer simulations, and field trips and teacher-centered strategies such as teacher demonstrations, and to review research on the effectiveness of various instructional methods in science.

STUDENT-CENTERED STRATEGIES

Inquiry Learning

Irrespective of the level at which students study science, they should receive an undistorted view of scientific activity, which implies an appropriate development of scientific skills and understanding of content. Inquiry learning, in the sense that it simulates real scientific activity, has a considerable role to play. As Woolnough and Allsop (1985) state:

Our goals for science teaching should be that students should acquire both a mature cognitive understanding of science knowledge and concepts and also mastery of the skills of a practicing scientist, which would involve both the development of basic laboratory skills and techniques and also development of the habit of working as a problem-solving scientist. [P. 71]

Welch et al. (1981) defined inquiry as "a general process by which human beings seek information or understanding. Broadly conceived, inquiry is a way of thought. Scientific inquiry, a subset of general inquiry, is concerned with the natural world and guided by certain beliefs and assumptions" (p. 33). More specifically, scientific inquiry is the process of conceiving problems, formulating hypotheses, designing experiments, gathering data, and drawing conclusions about a scientific problem or science phenomena. These abilities can be developed by involving students in activities requiring their performance. Through this type of learning and the acquisition and synthesis of scientific knowledge and processes, the ability to conduct scientific inquiry becomes possible.

In teacher-centered learning, students tend to be passive, with much of their activity consisting of listening and apprehending. Such an approach can allow little opportunity for developing inquiry skills. In contrast, inquiry learning places a major responsibility for the learning activity on the student. Welch et al. (1981) describe inquiry learning as generally associated with much involvement on the part of the student. The approach to learning provides firsthand experience of doing science. Students who learn by inquiry are responsible for developing their own answers to questions, rather than relying exclusively on the teacher and textbooks.

Inquiry teaching and learning frequently have been topics in the science education literature during the past quarter of a century. Welch et al. (1981) wrote that "the science education community has advocated the development of inquiry skills as an essential outcome of science instruction, and for an equal number of years science educators met with frustration and disappointment" (p. 1). In spite of new curricula, better trained teachers, and improved facilities and equipment, the optimistic expectation that students would become inquirers seldom has been fulfilled.

Although students in junior and senior high schools usually were taught observing and measuring skills, relatively few students were successful when tested for mastery of these skills (Welch et al., 1981). Moreover, if students were familiar with scientific phenomena, they were able to select hypotheses to explain them. If the situation or phenomenon was unfamiliar, however, their success in generating and selecting hypotheses decreased considerably. Welch et al. showed that proper inquiry teaching was used in only a very limited number of schools. Thus, as with other teaching strategies which potentially could be effective, evidence of application, rather than of success, still is being sought.

Science Laboratory Classes

For a long time the science laboratory has had a central and distinctive role in science education in involving students with concrete experiences of concepts and objects. In the 1960s, as part of the reform in science education, practical work in laboratories was supposed to be used to engage students in investigations, discoveries, and problem solving. The laboratory became the center of science education.

The laboratory is a unique educational setting in which students, usually in small groups, interact with materials and equipment and observe phenomena. These laboratory experiences can have different levels of structure specified by the teacher or laboratory handbook. They include four broad phases of activity: planning and design; performance; analysis and interpretation of results; and application (Kempa and Ward, 1975).

Goals of laboratory work. Laboratory activities traditionally have been used for a wide variety of cognitive, practical, and affective goals. In the cognitive domain, the goals would include promotion of intellectual development, enhancement of the learning of scientific concepts, development of problem-solving skills, development of creative thinking, and increased understanding of the methods of science. Practical goals would include development of skills in performing science investigations, in analyzing investigative data, in communicating, and in working with others. The affective goals would include enhanced attitudes toward science and positive perceptions of one's ability to understand and to affect one's environment (Giddings et al., 1991).

Although it is accepted intuitively that laboratory work is an essential part of science learning and that it provides a unique medium for teaching and learning in science education, little objective information supports the educational effectiveness of this instructional technique. In 1969, Ramsey and Howe wrote:

That the experience possible for students in the laboratory situation should be an integral part of any science course has come to have a wide acceptance in science teaching. What the best kinds of experiences are, however, and how these may be blended with more conventional classwork, has not been objectively evaluated to the extent that clear direction based on research is available for teachers. [P. 75]

On the basis of an extensive review of research on the educational effectiveness of the science laboratory, Bates (1978) wrote that "the answer has not yet been conclusively found: what does the laboratory accomplish that could not be accomplished as well by less expensive, less time-consuming alternatives?" (p. 75). More recently, Gallagher (1987) concluded that: "Laboratory work is an accepted part of science instruction. Given its important place in the education of youth, it is surprising that we know so little about its functioning and effects" (p. 301).

One of the causes for the lack of information about the effectiveness of the laboratory would seem to be that past research was not sufficiently comprehensive. It generally examined relatively narrow bands of laboratory-related skills, and the conclusions that were drawn applied to a narrow range of teaching techniques, teacher and student characteristics, and learning outcomes. Clearly, the school laboratory does have an important role to play in science education:

Theory and research suggest that meaningful learning is possible in laboratory activities if all students are provided with opportunities to manipulate equipment and materials while working cooperatively with peers in an environment in which they are free to pursue solutions to problems which interest them. [Tobin, 1990, p. 414]

Student and teacher behavior in the science laboratory are affected significantly by the nature of the laboratory activities. Considerable variation in the relative amount of responsibility assumed by the learner and the teacher is possible. Lunetta and Tamir (1979) analyzed laboratory handbooks and found great gaps between the stated goals for laboratory teaching and the kinds of activities that students generally are asked to perform. They found that, despite the curriculum reforms in the previous twenty-five years, students still commonly worked as technicians on "cookbook" activities requiring low-level skills and abilities. They were given few opportunities to discuss experimental error, to hypothesize and propose experimental tests, or to design and then actually perform an experiment. These large discrepancies between goals and practice are another important factor contributing to the ambiguous research findings on the educational effectiveness of the science laboratory. In other words, until the potential of the laboratory is exploited properly, we cannot expect to reach the benefits which intuition suggests must be there.

Teacher roles in laboratory work. Curriculum development of the 1960s showed that teachers play important roles in students' learning. The best curriculum materials can result in limited student growth if a teacher is insensitive to the intended goals, to student needs, and to appropriate teaching strategies. The teacher provides an organization and environment that affect whether or not students meet instructional goals. If a teacher's goal, for example, is to teach observational skills and not just facts that can be observed, this goal should be apparent in the things that the teacher says and does. Shymansky and Penick (1978) wrote:

Teachers are often confused about their role in instruction when students are engaged in hands-on activity. Many teachers are concerned about an adjustment they may have to make in their teaching style to facilitate a hands-on program as well as how students will react to increased responsibility and freedom. An activity-oriented classroom in which hands-on materials are made available to students is often a very new experience for the teacher as well as for his students. [P. 1]

Johnstone and Wham (1982) found that the amount of learning taking place in the laboratory is rather limited. A chief explanation is that students have to handle a vast amount of information which overloads their memories. Thus, laboratory exercises and inquiry should be organized and tailored to the needs and abilities of the students.

Assessment of students' laboratory performance. Laboratory activities can enable students to integrate their experiences with the materials and phenomena of science, conceptualize aspects of these activities, and construct formal schemes and models for practical investigation. Laboratory activities can involve both manual and intellectual abilities. These abilities are distinct from those used in work that is exclusively verbal. Laboratory skills can be classified in the following ways (Kempa and Ward, 1975):

1. planning and design of an investigation in which the student predicts results, formulates hypotheses, and designs procedures;
2. carrying out the experiment, which involves the student in making decisions about investigative techniques and manipulating materials and equipment;
3. observation of particular phenomena; and
4. analysis, application, and explanation of the findings.

Precisely which of these phases should be evaluated, of course, depends on the teacher's and researcher's pedagogical objectives and on the nature of the experiment. Kempa and Ward claim that these phases of experimental work are a valid and satisfactory framework for the development and assessment of both psychomotor and cognitive practical skills.

Bryce and Robertson (1985) reviewed the use of practical work in science education in different countries. They found that in many countries teachers spent (or claimed that they spent) a considerable amount of time in supervising laboratory work, yet the bulk of science assessment was traditionally nonpractical. In other words, the assessment of student performance in the science laboratory by and large is neglected in most countries and by most teachers.

Systems for evaluating student activity in the laboratory can be classified into four broad categories: written reports; test items; laboratory practical examinations; and continuous assessment (Giddings et al., 1991). Such evaluation is more possible today than in the past, because we now have the knowledge and the tools needed to make valid, reliable, useful, and challenging practical assessments in science.

We conclude this section with the relatively optimistic view expressed by Pickering (1980), which probably reflects the intuitive feeling of many science educators:

> The job of lab courses is to provide the experience of doing science. While that potential is rarely achieved, the obstacles are organizational and not inherent in laboratory teaching itself. That is fortunate, because reform is possible and reform is cheap. Massive amounts of money are not required to improve most programs; what is needed is more careful planning and precise thinking about educational objectives. By offering a genuine, unvarnished scientific experience, a lab course can make a student into a better observer, a more careful and precise thinker, and a more deliberate problem solver. And that is what education is all about. [P. 80]

Computer Simulations

The term "simulation" has been used in several ways in the natural and social sciences. To simulate means to imitate a real system (for example, an economy) or a process (like the flow of fluid through a pipe). Often the simulation operates on the basis of a mathematical or logical model. The model is intended to imitate the original faithfully, but generally includes fewer details (Lunetta and Hofstein, 1991). Instructional simulation enables learners to understand better the real system that is being simulated. Instructional simulations, as well as laboratory activities, enable students to interact with models of reality. Within contrived settings, both can enable students to confront and resolve problems, make decisions, and observe efforts.

The discussion here is limited to computer simulations that students use as an alternative and/or a complement to laboratory work. Even in inquiry-oriented laboratories, curriculum developers and teachers usually have to limit the scope of the activity, because of constraints of time, equipment, space, materials, and measurement error. Thus, computer simulations can provide an effective substitute or enhancement for many laboratory activities. Simulations can be planned to provide meaningful learning and to engage students in interactions with problems, models, or experiments that are too complex, dangerous, expensive, fast, slow, or time- and material-consuming. As a resource for inquiry, computers can assist in the collection, interpretation, and analysis of data; as an instrument in the laboratory, they provide instantaneous digital and graphic readouts of variables such as temperature, time, voltage, and velocity.

Effectiveness of computer simulations. Simulations can play important roles in science teaching and learning, but here again research has been limited. A small number of studies have compared the effects of laboratory simulations with more conventional laboratory work (Lunetta and Hofstein, 1991). These studies suggest, perhaps not surprisingly, that instructional simulations generally have not been as effective in promoting manipulative skills as are hands-on experiences. On the other hand, most studies showed that computer-simulated laboratories were at least as effective as conventional laboratory work for promoting concept learning as measured by paper-and-pencil tests (Eisenkraft, 1987). In general, simulation activities took considerably less time than activities in the laboratory, and students' attitudes tended to be positive toward both. The research typically reported learning outcomes on relatively narrow information and concepts, with few studies assessing important outcomes such as problem-solving skills, attitudes toward science, and understanding of scientific processes. Further research is needed to provide a sounder basis for such instructional practice.

Pedagogy of student inquiry, laboratory work, and computer simulations. In the school laboratory class students generally are assigned to small groups of two to six students for practical work. When used optimally, the small-group, cooperative learning environment enables students to share ideas, question each other, consider alternative ideas, and assist each other in understanding and problem solving. Small-group, cooperative learning can be identified by the following four elements: (a) positive interdependence; (b) face-to-face interaction; (c) individual accountability; and (d) the appropriate use of interpersonal and small-group skills (Slavin, 1983). Traditionally, students work independently and in competition with each other, and many do not know how to cooperate in achieving a group goal. The task of the teacher who wants to foster improved interpersonal skills is to structure the classroom and learning activities so that students have opportunities to use cooperative learning skills (Slavin, 1983).

Teaching strategies which promote cooperative learning can enhance cognitive outcomes, attitudes, social skills, and work habits (Slavin, 1983). Both laboratory activities and simulations provide opportunities in science teaching to engage students in such cooperative, small-group interactions and thus facilitate learning in science. Student success while learning in small groups might be attributed to relief from the excessive teacher-student interaction in whole-group teaching and to time freed for interactive engagement of students. The acts of tutoring and teaching in cooperative groups can encourage students to think for themselves about the organization of subject matter and productive time allocation.

Simulations (especially computer simulations) also can be used effectively by individual students, and some simulation software is designed so that the computer plays the role of a "lab partner." Simulations and laboratory activities also can be used, at times, for large-group demonstrations. They can be particularly appropriate when a teacher wishes to demonstrate how to conduct a special kind of inquiry before students work independently. Teachers also can use simulations or laboratory activities to engage an entire class in observing relationships and formulating hypotheses to explain observations. Ways in which to test those hypotheses then can be discussed in large or small groups, with the teacher being involved again when results are discussed toward the end of the activity. A simulation to facilitate postlaboratory discussion also can be helpful. As with all strategies, those mentioned here can be used deliberately to create planned effects and desired objectives, although spontaneity can be exploited by masterful teachers.

Field Trips/Outdoor Activities

Visits to farms, parks, school camps, nature centers, and other outdoor education enrichment settings are standard practices in education, and much has been written concerning their educational value (Koran and Baker, 1979). Most of what has been written, however, is based on intuition; surprisingly little research evidence about field trips is available. Do children learn from field trips? If so, what do they learn? What are the factors that influence what and how much they learn? How can schools use field trips more effectively? Only recently have such questions been investigated.

In contrast to conventional science classrooms, field trips take place in a more open environment, with fewer teacher sanctions and more flexible, potentially different evaluation procedures. The participants often are able to move around at their own pace and to explore on their own. Further, field trips can provide the student with concrete experiences unavailable in the classroom. Koran and Baker (1979) and Orion (1990) have shown that well-designed, planned, and structured field trips can provide an effective instructional experience. By providing an alternative to the normal science classroom setting, field trips can promote motivation and achievement. According to Koran and Baker (1979), field trips can be successful as an instructional strategy provided that: (a) the teacher is familiar with the area to be visited; (b) students are prepared and clear about the objective of the field trip; and (c) the field trip provides a meaningful experience that would not be available in the classroom.

One of the problems in obtaining evidence concerning the effectiveness of field trips is the method of evaluation used in past research. Most past studies have compared groups involved in field trips with groups not involved in field trips, using responses on traditional paper-and-pencil tests and questionnaires. Data obtained from such instruments fail to support definitive findings concerning achievement and attitudes. An alternative involves using case study or naturalistic inquiry research methods. For example, Orion (1990) observed classes and interviewed students and teachers in order to assess the use of field trips in the context of geological education in high schools in Israel. He obtained useful information concerning the planning, implementation, and conditions (for example, geographical and climatic factors) for educationally sound field trips. The use of alternative inquiry methods has the potential to yield a greater understanding of what students do when learning science while on field trips.

The study of instructional problems in informal settings has become an acceptable and productive line of research. Results so far seem to support the hypothesis that science teaching should include instructional experiences outside the conventional classroom, but that unsystematic and unplanned trips usually constitute unproductive and indefensible teaching strategies.

Distance Education

Distance education can free students from the limitations of space, time, and age, and has a record of success in both high- and low-income countries. It can include correspondence texts, books, newspaper supplements, posters, radio and television broadcasts, audio and video cassettes, films, computer-assisted learning, teleconferencing or video-conferencing, and self-instructional kits, as well as such local activities as supervised independent study, supplementary teaching, tutoring, counseling, and student self-help groups. Scarce resources of scientific, pedagogical, and media expertise concentrated in development centers for science education can be shared more widely using print and broadcast media. The use of books and other printed material, which ordinarily account for most of the cost of instructional materials, is a prime example of a strategy that has consistently worked better than exclusive use of oral teaching (Walberg, 1991).

Distance approaches can be highly cost effective when large numbers of students follow the same preproduced courses. Compared with a single teacher working alone, distance courses are more able to incorporate validated subject matter and systematic instructional design, and spread developmental costs over thousands of students. Such courses can provide for individualized study by including clear learning objectives, self-assessment materials, and opportunity for feedback either periodically or on demand.

Distance methods also can substitute for conventional science teaching in secondary schools. In "self-study" schools in China and Malawi, for example, radio or television provides instruction to classes supervised by an older student or adult. Also, distance methods can provide greater access and allow advanced subjects such as science to be taught in small day schools. Experience in sub-Saharan Africa suggests that rural students, especially girls, are more likely to attend small day schools close to their homes than distant boarding schools (Moock and Harbison, 1987). Because teacher knowledge and laboratory equipment can be in short supply locally, distance education can substitute for normal secondary school science. Distance methods also have been used successfully in high-density, high-income countries to enlarge opportunities for study, especially at the university level as the British Open University and the Chicago City Colleges have demonstrated (Walberg, 1991).

TEACHER-CENTERED STRATEGIES

Conventional Teaching

Direct or conventional teaching has predominated throughout the history of universal education. The National Science Foundation studies

(Wise, 1978), conducted in the late 1970s in the United States, indicated that the dominant methods of teaching were the lecture and recitation, with the teacher being in control. The key to the new information was the textbook. In any classroom discussion, most of the questions asked were posed by the teacher and were low-level factual questions. Ideally, the teacher-centered lecture, however, can be educationally sound provided that teachers (a) ask questions that require students to demonstrate the ability to comprehend, apply, and analyze, and (b) give students reasonable time to respond to questions and wait longer before they act (that is, more "wait time").

Rowe (1973) showed that science teachers not only ask too many questions, but they also tend to ask them far too rapidly: "When wait times are short and reward schedules are high, payoff for students comes in doing only one thing . . . focusing totally on the wants of the teacher" (p. 243). In a meta-analysis (Wise and Okey, 1983) concerning the effect of instructional techniques on science achievement, wait time was found to be the most powerful technique employed by teachers to increase students' cognitive outcomes, critical thinking, creative thinking, and affect. Support for this finding also is given by Fraser et al. (1987) in a synthesis of research on educational productivity.

Teacher Demonstration

One of the most frequently observed activities in science classrooms in the USA is the teacher demonstration, which has been found to be conducted at least once a week in two out of five classes (Wise, 1978). Demonstrations can be conducted deductively, with the student involved in verification and observation, or inductively, with the student involved in high-level inquiry skills. Well-planned inductive demonstrations provide students with a stimulus to think and with immediate feedback from the teacher. The feedback acts as a guide for further questioning until students discover the principles involved in the demonstration. Whether conducted by the teacher or presented by another vicarious technique such as films or videotapes, demonstrations can be justified as an alternative to student laboratory work. This is especially so in situations in which, as a result of various factors such as cost of equipment, time, and safety, a truly student-centered laboratory approach is not feasible. There is much to be said for a blend of practical work by students and teacher demonstrations because past research clearly has shown that well-planned teacher demonstrations are as effective as students' personal experimentation (Bates, 1978).

THE TEACHER'S ROLE IN DIFFERENT INSTRUCTIONAL STRATEGIES

Table 4-1 summarizes the teacher's roles associated with the various instructional strategies discussed in this chapter. It is clear from the table that the more the instructional strategy becomes student-centered, the more the "center of gravity" of the teacher activity moves to the right of the table (that is, to activities which, as experience has shown, are much more demanding on the teacher). For example, the teacher must be prepared to face unforeseen and unplanned questions and phenomena posed by the students. This requires that teachers be open-minded, tolerant, and flexible

Table 4-1

Teacher's Roles in Different Instructional Techniques

Instructional Strategy	Teacher's Roles					
	Lecturing	Providing Information	Demon-strating	Managing	Guiding and Facilitating	Helping to Analyze Data and Results
Conventional teaching	+	+	(+)			
Demonstration	+		+			(+)
Classroom discussion		+	(+)	+		+
Laboratory class				+	+	+
Group learning		+			+	+
Inquiry		+			+	+
Field trip				+	+	+
Computer simulation			+		+	+
Individual learning		+			+	+

(+) Possible teacher role

and that they encourage students' ideas, questions, and comments in order to promote critical and analytical thinking. Student-centered instructional strategies can increase not only student activity, but also teacher activity.

RESEARCH FINDINGS ON EFFECTIVENESS OF INSTRUCTIONAL STRATEGIES

In this section we review some of the general research findings relevant to the effectiveness of instructional strategies. In recent years, Walberg and others have compiled syntheses of thousands of research studies in education (Fraser et al., 1987; Walberg, 1991). These syntheses suggest that the following nine constructs, classified here under student aptitude, instruction, and psychological environments, appear to be associated with increased learning:

Student aptitude: ability or prior achievement; development as indexed by chronological age or stage of maturation; motivation or self-concept as indicated by personality tests or the student's willingness to persevere intensively on learning tasks.

Instruction: amount of time that students are engaged in learning; the quality of the instructional experience including method (psychological) and curricular (content) aspects.

Psychological environment: the "curriculum of the home"; the morale or climate of the classroom or school group; the peer group outside the school; the mass media, especially the amount of television viewing in leisure time.

Using synthesis techniques, estimates of the size of the contributions of each construct to general learning outcomes have been compiled. In particular, it was found that ability or prior achievement, development, motivation, and amount and quality of instruction are essential for learning in school.

Quality of instruction is the main factor of direct relevance to us here. It can be understood abstractly as providing optimal cues, correctives, and encouragement to ensure the fruitfulness of engaged time. For example, diagnosis and tutoring can help to insure that instruction is suitable for the individual student, and inspired teaching can enhance motivation and thus keep up students' interest and perseverance. Improving the quality of instruction, then, can be considered as the efficient enhancement of learning time.

The status of instructional strategies in science education was investigated in two major meta-analyses by Wise and Okey (1983). On the whole, it was found that the size of the effects for most strategies was moderate except for wait time, for which the effect was large although the results were based on very few studies.

Table 4-2 summarizes research information on the effectiveness of various instructional behaviors (based on Wise and Okey, 1983), and clearly reflects the strength of wait time as a determinant of student achievement. Wise and Okey offer the hope that combining effective strategies might yield larger effects.

Table 4-2

Science Teacher Behaviors: Mean Effect Sizes

Behavior	Mean Effect Size*	Standard Deviation	Number of Studies
Wait-time	0.90	0.43	4
Focusing (e.g., organizers)	0.57	0.91	28
Manipulative (by students)	0.57	0.64	24
Modified (specific content)	0.52	0.45	22
Questioning	0.48	0.39	13
Inquiry-discovery	0.32	0.73	58
Testing (e.g., diagnostic)	0.32	0.46	45
Presentation mode (e.g., team teaching)	0.26	0.56	103
Teacher direction (extent of)	0.23	0.66	45
Audiovisual methods	0.18	0.48	33
Grading (e.g., pass-fail)	-0.15	0.38	14
Miscellaneous	0.43	0.26	12
Overall mean = 0.34		Total = 411	

Based on Wise & Okey (1983)

* The effect size is the number of standard deviations by which the mean of the experimental group exceeds the mean of the control group.

MATCHING INSTRUCTION TO
STUDENT MOTIVATION

A widely accepted notion is that instructional techniques should be matched to learners' characteristics and needs if the effectiveness of the teaching/learning process is to be maximized. Learners' characteristics that have received attention include cognitive characteristics such as achievement, cognitive level, conceptual level, and certain affective traits such as attitudes, interests, and motivation (reviewed by Kempa and Diaz, 1990).

Hofstein and Kempa (1985) suggested that students have preferences for particular types of learning activities and that these reflect their motivation. Adar (1969) identified four motivational "needs": the need to achieve; the need to satisfy curiosity; the need to discharge a duty; and the need to affiliate with other people. The types of learner corresponding to these needs were referred to by Adar as "achiever students," "curious students," "conscientious students," and "sociable students." After studying the relationship between students' motivational needs and their preference for particular teaching and learning strategies, Adar concluded that "the application of different teaching techniques will affect a student's motivation only if the method interacts with the student motivational pattern."

Hofstein and Kempa (1985) elaborated this theory for science education. They postulated that a number of relationships exist between students' motivational needs and their preferences for particular modes of instruction in science education. These attempts to relate the most significant instructional features associated with teaching/learning to the learner's motivational pattern are summarized in table 4-3.

Students have different motivational traits which lead to different preferences or dislikes for certain instructional strategies used in science education (Kempa and Diaz, 1990). This finding should be taken into consideration in the design and planning of learning experiences and teaching interventions, both by teachers and curriculum developers, as well as those who are responsible for the organization of science education in schools. Clearly, it is difficult in practice to respond to each student's needs, but much could be achieved if teachers would use a wide repertoire of instructional strategies instead of limiting themselves to only one or two. To do so requires not only skill in a variety of teaching methods but

Table 4-3

**Relating Instructional Features to Students'
Motivational Characteristics**

Type of Activity	Examples	Comment on Suitability/Unsuitability
Discovery/inquiry-oriented learning methods Problem-solving	Advocated in many science programs developed in the USA and UK during the 1960s and 1970s	Suitable mainly for students with "curiosity"-type motivational pattern Insofar as problem-solving activities are likely to require students to engage in judgment and evaluation situations (both tend to involve "high risk" taking), these are disliked by both 'achievers' and 'conscientious' students.
Open-ended learning activities (student-centered)	Learning activities without clearly specifiable objectives, except those relating to scientific processes (i.e., those associated with project work or student research)	Strongly preferred by the 'curious', but not other motivational groups which prefer clear teacher direction regarding educational goals.
Formal teaching with emphasis on information and skill transfer	Conventional 'traditional' instructional procedures, involving frontal teaching (e.g., with clearly defined goals and objectives)	Preferred by 'achievers' and 'conscientious' students because only low level of risk-taking is needed.
Collaborative learning activities	Games, simulations	The majority of games and simulation exercises devised for science education are "interactive' and, hence, particularly suitable for learners with a strong social motivation pattern. However, 'achievers' are likely to be opposed to an involvement in this type of learning activity.

also managerial competence to implement them in the classroom. This clearly represents a major challenge to in-service education for teachers and teacher educators.

CONCLUDING REMARKS

Our purpose in this chapter has been to identify various student-centered and teacher-centered instructional strategies and to review research evidence regarding their effectiveness. Research suggests that some instructional techniques can help students attain some of the goals of science learning. Based on the evidence presented here, however, it would be unreasonable to expect that any single method will be effective for achieving all goals in science education.

Appropriate instructional activities can be effective in promoting the development of logical thinking, as well as the development of some inquiry and problem-solving skills. Some instructional techniques promote manipulative and observational skills and understanding of scientific concepts. Some foster cooperation and communication. More research into the enhancement of the application and orchestration of these strategies is needed.

Finally, based on the research reviewed here, the following implications for improving science education can be drawn:

1. In order to increase students' motivation to learn science, a variety of instructional techniques should be used.
2. Well-designed demonstrations by the teacher can be effective substitutes for students' own laboratory work.
3. The laboratory is a unique educational setting in the sciences which can be used to pursue certain goals such as interest in science, motivation, and manipulative skills.
4. Alternative instructional methods such as computer simulations and teacher demonstrations are more efficient than laboratory work for achieving some educational goals when there are problems of time, space, lack of equipment, and safety in using laboratories.
5. Student outcomes should be assessed thoroughly both in the classroom and in the laboratory.
6. Laboratory activities and computer simulations provide opportunities in science teaching to engage students in small-group cooperative interactions.
7. Field trips provide students with more freedom and flexibility and when well designed, planned, and structured can provide an effective learning experience.

8. Distance education is an effective method in high-income countries to enlarge opportunities for study. In low-income countries, it could be used as a substitute for secondary schooling in cases in which there are limitations of space, time, and money.

9. Because one of the key factors associated with learning is "wait time," there is a need to train teachers in using this technique.

REFERENCES

Adar, Lea. *A Theoretical Framework for the Study of Motivation.* Jerusalem, Israel: Hebrew University, 1969. In Hebrew.

Bates, Garry R. "The Role of the Laboratory in Secondary School Science Programs." In *What Research Says to the Science Teacher,* vol. I, edited by Mary Budd Rowe, pp. 55-82. Washington, D.C.: National Science Teachers' Association, 1978.

Bryce, Tom G. K., and Robertson, Isobel J. "What Can They Do? A Review of Practical Assessment in Science," *Studies in Science Education* 12 (1985): 1-24.

Eisenkraft, Ahron J. "The Effect of Computer Simulated Experiments and Traditional Laboratory Experiments on Subsequent Transfer Tasks in a High School Physics Course," *Dissertation Abstracts International* 47 (1987): 3723.

Fraser, Barry J.; Walberg, Herbert J.; Welch, Wayne W.; and Hattie, John A. "Syntheses of Educational Productivity Research," *International Journal of Educational Research* 11 (1987): 73-145 (whole issue).

Gallagher, James J. "A Summary of Research in Science Education," *Science Education* 71 (1987): 244-284.

Giddings, Geoffrey J.; Hofstein, Avi; and Lunetta, Vincent N. "Assessment and Evaluation in the Science Laboratory." In *Practical Science,* edited by Brian Woolnough, pp. 167-178. Milton Keynes, England: Open University Press, 1991.

Hofstein, Avi, and Kempa, Richard F. "Motivating Strategies in Science Education: Attempt at an Analysis," *European Journal of Science Education* 7 (1985): 221-229.

Holdzkom, David, and Lutz, Pamela M. *Research Within Reach: Science Education.* Washington, D.C.: National Science Teachers' Association, 1989.

Johnstone, Alex H., and Wham, A. J. B. "The Demands of Practical Work," *Education in Chemistry,* May (1982): 71-73.

Kempa, Richard F., and Diáz, Maria M. "Motivational Traits and Preferences for Different Instructional Modes in Science Education: Part I," *International Journal of Science Education* 12 (1990): 195-205.

Kempa, Richard F., and Ward, John E. "The Effect of Different Modes of Task Orientation: An Observational Attainment in Practical Chemistry," *Journal of Research in Science Teaching* 12 (1975): 69-76.

Koran, John J., and Baker, Dale. "Evaluating the Effectiveness of Field Experiences." In *What Research Says to the Science Teacher,* vol. II, edited by Mary Budd Rowe, pp. 50-67. Washington, D.C.: National Science Teachers' Association, 1979.

Lunetta, Vincent N., and Hofstein, Avi. "Simulation and Laboratory Practical Activity." In *Practical Science*, edited by Brian Woolnough, pp. 125-137. London, England: Open University Press, 1991.

Lunetta, Vincent N., and Tamir, Pinchas. "Matching Laboratory Activities with Teaching Goals and Practice," *Science Teacher* 46 (1979): 22-24.

Moock, Peter R., and Harbison, Frank W. *Education Policies for Sub-Saharan Africa: Adjustment, Revitalization, and Expansion*. Washington, D.C.: Population and Human Resources Department, World Bank, 1987.

Orion, Nir. "Educational Field Trips in High School Geology: Development, Implementation and Evaluation." Ph.D. dissertation, Weizmann Institute of Science, Rehovot, Israel, 1990.

Pickering, Miles. "Are Laboratory Courses a Waste of Time?," *Chronicle of Higher Education* 19 (1980): 80.

Ramsey, Gregor A., and Howe, Robert W. "An Analysis of Research on Instructional Procedures in Secondary School Science (II)," *Science Teacher* 36 (1969): 72-81.

Rowe, Mary Budd. *Teaching Science as Continuous Inquiry*. New York, N.Y.: McGraw-Hill, 1973.

Shymansky, James A., and Penick, John E. "Teacher's Behavior Does Make a Difference in Hands on Science Classroom." Unpublished report. Iowa City, Iowa: Department of Science Education, University of Iowa, 1978.

Slavin, Robert E. *Cooperative Learning*. New York: Longman, 1983.

Tobin, Kenneth. "Research on Science Laboratory Activities: In Pursuit of Better Questions and Answers to Improve Learning," *School Science and Mathematics* 90 (1990): 404-418.

Walberg, Herbert J. "Improving School Science in Advanced and Developing Countries," *Review of Educational Research* 61 (1991): 25-69.

Welch, Wayne W.; Klopfer, Leopold E.; Aikenhead, Glen S.; and Robinson, James T. "The Role of Inquiry in Science Education: Analysis and Recommendations," *Science Education* 65 (1981): 33-50.

Wise, Iris R. *National Survey of Science, Mathematics and Social Studies Educational Practices in U.S. Schools: An Overview and Summaries of Three Studies*. Washington, D.C.: National Science Foundation, 1978.

Wise, Kevin C., and Okey, James R. "A Meta Analysis of the Effects of Various Science Teaching Strategies on Achievement," *Journal of Research in Science Teaching* 20 (1983): 419-435.

Woolnough, Brian, and Allsop, Terry. *Practical Work in Science*. Cambridge, England: Cambridge University Press, 1985.

Chapter 5

STUDENT ASSESSMENT AND CURRICULUM EVALUATION

Wayne W. Welch

This chapter is about two related but different concepts: assessment and evaluation. Assessment is a term that has grown in prominence in recent years due in part to the appearance and popularity of needs assessment, national and international assessments of educational progress, and assessment of students' disabilities to determine eligibility for mandated compensatory programs. "Assessment" often is used interchangeably with testing and measurement, and is closely related to grading or marking. However, in many places testing has become closely aligned with group-administered paper-and-pencil exercises, often multiple-choice in format, while measurement mainly is associated with the quantifiable aspects of behavior that lend themselves to statistical analysis. Marking involves assigning a number or letter to a student based upon the student's achievement on examinations given by the school (internal) or administered by a central authority (external). These marks are used by schools and parents as indicators of students' success. Assessment, as used in this chapter, describes a broader array of procedures that are being used to gather information about what a person knows, feels, or can do.

The context for this discussion of assessment and evaluation is the push for the improvement of science education for students at the elementary and high school levels. Several major developments in assessment are described and their impact on science education discussed.

Chapter consultants: Pinchas Tamir (Hebrew University of Jerusalem, Israel), Campbell McRobbie (Queensland University of Technology, Australia), and Paul Black (University of London, England).

The general testing of students for the purpose of assigning grades or marks is not included in this chapter. However, the interested reader is referred to Tamir et al. (1987) for a discussion of the marking process in Australia, Denmark, the Netherlands, Italy, Scotland, the United States, and Israel.

Evaluation often is defined as the systematic investigation of the worth or merit of some object. Evaluation involves the gathering of information, as does assessment or measurement, but the distinguishing characteristic of evaluation is found in its value emphasis. That is, evaluation is concerned with the merit or worth of objects (for example, programs, processes, or materials) as determined from information gathered about the objects. Curriculum evaluation, then, is the determination of the merit or worth of a curriculum.

In this chapter, I describe and analyze the nature of recent evaluation activity in science by focusing on the extent and nature of (a) formative evaluation, (b) summative evaluation, and (c) curriculum research conducted for the improvement of science education. I first examine recent developments in student assessment. This is followed by a description and analysis of curriculum evaluation efforts. Finally, I summarize the major findings and propose several recommendations.

STUDENT ASSESSMENT IN SCIENCE

Prompted in part by massive federal efforts to reform the preuniversity science curriculum and a growing interest in educational accountability, the field of science assessment has undergone several changes during the past twenty years. The reform efforts, initiated in the late 1960s and early 1970s, were directed at modernizing the science curriculum, improving science literacy, and producing sufficient numbers of scientists to meet the needs of an increasingly technological society. Changes in the curriculum required changes in the nature of science tests. One of these changes was a different view of what should be tested.

Demands for educational accountability (that is, assigning to schools the responsibility of providing evidence that they are meeting their goals) and the growth of program evaluation also exerted considerable influence on the nature of science assessment. Transformations in the uses and purposes of science tests have occurred. The nearly exclusive use for evaluating student progress has been replaced by an expanded role which includes program evaluation, school accountability, state, national, and international assessments, and the improvement of instruction. Changes in the nature and purposes of science testing are discussed in the following sections.

Changes in the Nature of Science Assessment

The new science curricula differed from their predecessors in several ways. The content was modernized, most programs were oriented to an overriding conceptual scheme, and variety and flexibility were stressed. Greater attention was given to the assessment of attitudes and to the nature of scientific inquiry. The unifying themes of the first generation of projects were science oriented, as illustrated by examples of chemical bonding (the Chemical Bond Approach) or physics (the Nuffield Physics Project). However, later groups shifted to themes that reflected an increasing emphasis on the interactions among science, technology, and society (for example, the applied chemistry program in Israel and the Man-Made World in the United States).

Each curriculum group developed a variety of learning aids including texts for students, teachers' guides, laboratory manuals and equipment, films, and tests. Most extolled the virtues of "hands-on" experiences and many opportunities were provided for the doing of science rather than reading about science. Developing skills in problem-solving, in inquiry, or in the processes of science was part of the objectives of all the new courses.

The tests developed for these new projects reflected their broader goals and attempted to measure the higher-level cognitive abilities. For example, the Biological Science Curriculum Study (BSCS) developed its own paper-and-pencil test on science processes for measuring students' performance and for evaluating the effectiveness of the curriculum (Welch, 1979). An evaluation of the Australian Science Education Project (ASEP) involved the development and use of measures of students' attitudes toward science and of classroom learning environment, as this curriculum sought to develop in students some understanding of the nature, scope, and limitations of science, as well as the skills and attitudes important for scientific investigation (Fraser and Cohen, 1989).

Although few of these earlier curriculum projects have survived as originally intended, their goals and objectives have been incorporated into commercially available textbooks. They also have influenced the character of our present national and international assessments. For example, the measurement of students' attitudes was a major focus of a science assessment for the first time in the United States in 1976-1977 and was included in the National Assessment of Educational Progress in 1981-1982 and in 1985-1986. Furthermore, paper-and-pencil items measuring inquiry or process skills now are part of national assessment in the United States and the United Kingdom and are integral parts of the IEA (International Association for the Evaluation of Educational Achievement) assessments in science (1970, 1983, and 1994/1995). They also were included in two international studies recently conducted by the Educational Testing Service, a large American testing company (IAEP, 1989).

A concern of science educators about most standardized testing programs is their limited capacity for assessing the "practical" performance skills which students are expected to learn in the science class or laboratory. Many of the curriculum projects, both past and those currently being developed, plan for students to spend up to half of their time on "hands-on" experiences. The expectation is that students will learn the skills of scientific investigation. However, the cost of testing for these skills on a national basis has been prohibitively high, and problems have occurred with administration of the tests and with the reliability of scoring.

Notable exceptions to this situation have occurred in Israel and the United Kingdom where practical examinations have been used for some time. In addition, the IEA Second International Science Study (SISS) included performance items for the first time on an optional basis in its 1983 assessment. Six countries (Hungary, Japan, Israel, Korea, Singapore, and the United States) chose to participate in this phase of the study.

Typically, students are given a set of materials and asked to use them to solve a problem. Some sample questions from these "practical tests" include the following:

1. Using the materials provided, connect the battery, switches, and bulbs so that the light will burn.
2. Make a saturated solution of ammonium chloride in a test tube and note the lowest temperature reached.
3. Find the factors that influence the period of a pendulum.
4. Using an iodine solution and sugar test strips, test for the presence of sugar and starch in three unknown solutions.

In general, these practical examinations are time-consuming and costly, and scores on them have rather low interrater reliability. However, because they represent so vividly the essence of science, they are receiving increased support in the science education community. For example, with support of the National Science Foundation, the Educational Testing Service carried out a feasibility study for including practical items in national assessments. Drawing on the work of the Assessment of Performance Unit (APU) in the United Kingdom, a manual for hands-on testing was developed for use by teachers (NAEP, 1987).

The growing interest in alternative assessment techniques has led to further developments in the United States at the national level and in a number of states (for example, California, Connecticut, Vermont, and New York). Alternative or "authentic" assessment techniques include the use of interviews, portfolios, simulations, concept mapping, and laboratory performance measures (Kulm and Malcom, 1991). Work continues as well in Australia, Israel, and the United Kingdom, although some opposition to this kind of testing has occurred due to such factors as cost, politics, problems with managing the process, and comparability of results (Black, 1993).

As indicated above, another development is the increased attention being given to the assessment of students' attitudes in science. The 1977 and 1982 national assessments in science (Hueftle, Rakow, and Welch, 1983) devoted approximately 20 percent of their testing time to attitude assessment. Students were asked, among other things, about their opinions of science classes, science teachers, careers in science, their involvement in science activities, and support of science research.

An example of the kind of attitude items used in these assessments is a set of questions about science teachers. Student opinion toward their science teachers was obtained from a national random sample of 2,000 17-year-olds. Results were reported in terms of the percentages of students expressing favorable responses. For example, in 1982, 49 percent of the respondents agreed or strongly agreed that their teacher made science exciting, while 63 percent thought that their teacher was enthusiastic. Results also were compared across time; for example, the latter response represented an increase of 5 percent since the 1977 assessment.

The 1986 national assessment in the United States also included attitude items. Seven categories of questions were asked: attitude toward science classes, careers, socio-scientific responsibility, science as a personal tool, value of science, societal issues, and student experience in science. About one eighth of the assessment addressed the affective domain.

Changes in the Purposes of Science Tests

For most of this century and in most countries, testing has been viewed primarily as a process for determining student progress. This information was used by schools and colleges to determine whether or not students should be promoted, graduated, or admitted to the next higher level of educational opportunity. In recent years, this gatekeeping function of testing has been modified to include the use of tests for making decisions about program quality and in research on science testing aimed at improving the process of instruction. The program or policy function of assessment is considered in the following section. Changes in science assessment for improving instruction and learning are discussed below.

Policy focus. By the middle of the century, two major testing movements were well established: (a) standardized testing in the United States, some provinces in Canada, and in several states in Australia; and (b) the external or school-leaving examinations, such as those found in England or Scotland. Both tests were initiated for the same general purpose, namely, to assist colleges and universities in deciding who should be admitted to higher education. A major difference is that standardized tests are not linked directly to the school curriculum, whereas external examinations influence the curriculum. Both forms of testing include science among the various subjects assessed, although it is usually an optional part of the examination.

The first issue in 1938 of Buros's *Mental Measurements Yearbook*, the standard reference for test users, listed thirty different science tests. In the eighth Yearbook in 1978 (Buros, 1978), the number of science tests had risen to forty-four, with the majority of them being included in the batteries of tests published by some of the larger companies (for example, the *Sequential Tests of Educational Progress* produced by Educational Testing Service and published by Addison-Wesley). The expressed purpose of these tests is to facilitate decisions about students' admission, placement, promotion, and graduation. Millions of students take these tests each year and several of the commercial test publishers reap substantial profits from the extensive testing that is done.

In recent years, growing criticism has been leveled at standardized tests including the limitations of multiple-choice formats, the pigeon-holing effects of test scores, bias against girls, minorities, and low-income groups, and low predictive validity. In addition, science tests have been criticized for their failure to represent the content of the newer science curricula (Darling-Hammond and Lieberman, 1992; Welch, 1979).

Simultaneous with the rising criticism of standardized testing was a growing interest in educational accountability. Institutions and curricula were being asked to provide evidence that they were meeting their responsibilities. As the demands grew for accountability and its hand-maiden, program evaluation, the role of testing underwent a shift in focus. No longer was the student the only focus of the testing activity. Educational institutions and programs were being examined as well. A new philosophy of testing developed that addressed questions regarding program effectiveness. For example, is curriculum A more effective than curriculum B? To what extent is the science program at Lincoln School meeting its objectives? What is the level of science literacy in the state of Virginia? How are students in the United States performing in science compared to students in other countries?

Policy questions of this sort have given rise to a number of new test-ing programs, including state, national, and international assessments, and mandated program review and evaluation at the local level. Assess-ing individual performance has been joined by measuring the perfor-mance of groups of students in order to draw inferences about program effectiveness. Testing becomes program-centered rather than student-centered. Many states and school districts require a periodic review or examination process that yields information about program effective-ness. Low scores are considered as indicative of possible problems in the system rather than indicators of students' failure.

An example illustrates this policy use of testing in addition to provid-ing insights into the developments in evaluation, a topic addressed in the second part of this chapter. A national curriculum evaluation was made of a new physics course called Harvard Project Physics (Welch, 1973). Physics teachers were selected randomly from throughout the country and assigned randomly to teach the new course or to continue teaching their former course. Students were tested before and after their respective courses using a variety of instruments. Physics achievement was assessed together with knowledge of science processes, attitudes toward various aspects of physics, and the socio-psychological environment of the physics class. Comparisons were made between the change scores in Harvard Project Physics classes and those in comparison classes. The findings were used to help inform marketing, accountability, adoption, and adaption decisions.

A policy role for testing perhaps is most apparent in the state, national, and international assessments that have been conducted during the past twenty years (Welch, in press). In 1969-1970, the first International Association for the Evaluation of Educational Achievement (IEA) and the first National Assessment of Educational Progress (NAEP) in the United States were conducted. Both groups chose science for their beginning efforts. In 1974, the Department of Education and Science in England initiated a national assessment program with science as one of four areas being surveyed.

In the United States the National Assessment is an ongoing congressionally mandated study to determine the nation's progress in education. Established in 1969, it has periodically gathered information on the educational achievement of students in reading, writing, mathematics, science, and social studies. Occasionally, it has evaluated student performance in literature, art, music, citizenship, computer competence, and career and occupational development. Reading performance and achievement in science have been assessed most often during the life of NAEP. Science was assessed in 1970, 1973, 1977, 1982, 1986, 1990, and 1994.

Until 1980, NAEP conducted annual assessments of 9-, 13-, and 17-year-olds attending public and private schools. In recent years, biennial assessments have been carried out and the students selected for testing came from grades 4, 8, and 12 rather than from age groups. Groups of items were assembled into a booklet which then was administered to national random samples of about 2,000 students. Approximately ten booklets were used at each age or grade level, yielding students' scores on about 300 items per group. About half of the items are repeated during consecutive assessments so that changes in national performance can be monitored. Results are reported for the country and for various subgroups (for example, gender, race, region of country, and type of community).

IEA studies are similar in format and purpose to the NAEP studies except that students from several different countries are involved. An elaborate committee structure is used to analyze the curriculum of each participating country, define the domain of content to be tested, and write items. In the first science assessment, nineteen countries participated. The second assessment in the mid-1980s involved twenty-four countries. Testing is done by age level (called "populations,") and by subject area (for example, chemistry, physics, earth science, and biology). (A detailed discussion of the IEA science studies can be found in chapter 8 of this book.) A third IEA study in science, the Third International Mathematics and Science Study (TIMSS), is currently in progress in more than forty countries (Robitaille et al., 1993).

Another international assessment study in science and mathematics was introduced in 1986 involving six different countries. It was called the International Assessment of Educational Progress (IEAP) and was carried out by the Educational Testing Service. A second implementation of this science and mathematics assessment was conducted in 1992 in approximately twenty countries. IAEP is unique in that it uses existing items written for NAEP instead of writing its own items. Furthermore, only 13-year-olds have been tested. Other assessment procedures are similar to those used in NAEP or IEA, but much less attention is devoted to interpreting and explaining findings than in the IEA studies.

There are some differences between the procedures used for the program-centered assessments described above and student-centered assessment. Perhaps most important is the use of randomized data collection or "matrix sampling" in program evaluation. Not every student needs to take every item as is required when decisions need to be made about students. A random sample of students responds to a random sample of items. Individual student scores are not reported; rather, groups of scores are combined to make estimates of school, state, or even national means.

Another difference in the testing process is the manner in which items are chosen. In traditional test development, items are selected because of their ability to discriminate among students with different achievement levels. However, in large-scale assessments, items are chosen according to their match to the domain of content defined for a particular subject such as science. In the language of test developers, such assessments are domain-referenced rather than norm-referenced.

As one might expect, the manner of reporting results is quite different in these program-centered assessments. The average response of a group to individual items or sets of items typically is used rather than reporting the scores of individuals. For example, it is common for international assessments to report their findings by stating that 9-year-olds averaged 60 percent on twelve items measuring knowledge of physical science. Furthermore, when tests are used to evaluate curriculum effects, one might read that students in the Israeli biology program scored half a standard deviation higher than students in a traditional biology course.

Because the average percentage of students correct on an item intrinsically is not very revealing, it is common to compare the responses of different groups (for example, age levels, countries, genders, ethnic backgrounds, or changes across time). The formative evaluation of the Australian Science Education Project (ASEP) illustrates this approach (Fraser and Cohen, 1989). Pretests and posttests were administered to students studying an ASEP unit and to a control group. Individual item means were calculated before and after instruction to determine if the unit was successful in promoting achievement of specific intended aims.

Another illustration of reporting assessment results is provided by the first IAEP assessment in science among 13-year-olds (IEAP, 1989). Average proficiency levels were reported for each country, except for Canada for which the median score of the seven groups assessed in Canada is reported. (Several different provinces were tested and, in some cases, examinations were given in English or French.) Item-response theory techniques were used to define proficiency for the set of items used in the assessment. This analysis provides a common scale on which performance can be compared across different groups (for example, age level, country, year of assessment). The policy implications of such studies seem rather obvious, particularly for those countries where proficiency levels are judged to be unsatisfactory.

Instructional Research Focus. Much research in science education in recent years has used clinical or in-depth interview techniques to study students' abilities, measure science learning, and investigate the effects of prior knowledge on learning. Much of this research has followed the intellectual development model of Piaget and reviews of research usually have a section devoted exclusively to such studies. For example, the chapter on teaching the natural sciences (White and Tisher, 1986) in the third edition of the prestigious *Handbook of Research on Teaching* contains more than seventy-five references to Piagetian studies.

In a typical research task two cups of the same size and same shape are filled with equal amounts of water. A child is asked to judge whether they contain the same amount of water. The water from one cup then is poured into a third cup which is taller and narrower than the first. The child then is asked to compare the amount in the narrow cup with the original untouched cup. Although there is a change in the shape and location of the water, the amount of water does not change. However, many children believe that the taller cup contains more water. Understanding of this conservation principle is thought to be related to the child's level of cognitive development. Knowing the operational level of the child helps to inform subsequent science instruction.

Other Piagetian tasks seek to measure understanding of conservation of length and weight, classification, transitivity, one-to-one correspondence, proportional reasoning, separation of variables, and the like. Much of the work has focused upon measuring the level of success achieved on the various tasks and then correlating this success with other variables such as instructional strategies or science achievement.

The in-depth interviews used to probe the intellectual development of children are time-consuming and require considerable interviewing skill. Therefore, paper-and-pencil group tests have been developed that try to measure similar traits. However, because correlations between scores on these tests and interview techniques have not been consistently high, some people have concluded that such tests are measuring different things (White and Tisher, 1986).

Although Piagetian research is still popular, the last several years have witnessed an increasing interest in another use of clinical interviews in science assessment. A common term used to describe this work is "student misconceptions," although a number of researchers prefer to speak of the "alternative frameworks" held by children and avoid the pejorative "misconceptions" term. Chapter 3 of this volume provides a comprehensive overview of methods and findings for studies of children's science conceptions and thinking.

This work is related to a constructionist philosophy which holds that children, as a result of their life experiences, construct a view of the world that seems to make sense to them. These views are variously called "life-world views," "misconceptions," "children's science," "mental structures," or "alternative frameworks." Unfortunately for educators, the child's view is often different from the scientist's view of the world, thus creating a conceptual gap that has implications for the teaching and learning of science. These gaps are thought to be particularly wide for concepts that are described by words that are common both to children and scientists, but that are used in different ways. Some examples include heat, temperature, energy, force, current, work, plant, animal, living, and conservation.

The alternative frameworks of children are thought to be consistent, and are quite resistant to change. However, they should be accepted and recognized by teachers. Children are able to recognize scientific views, but they have difficulty incorporating them into their own views once they leave the classroom. Much of the current research is devoted to identifying and assessing the range of misconceptions in science held by children and, in many cases, by adults.

An example illustrates the nature of this research. A child is presented with two cups containing equal amounts of water. They are both at the same temperature of, say, 10 degrees Celsius. The two cups then are poured into a third cup and the following question is asked: "What is the temperature of the mixture?" In addition, the children are asked to explain the reasons for their answers.

The majority of children up to the age of 13 years will report that the new temperature is 20 degrees instead of the correct answer of 10 degrees (Strauss and Stavy, 1983). Among some age groups (for example, 8-year-olds), only 30 percent of children get the correct answer. There is confusion between temperature, a measure of the average kinetic energy, and heat, the total energy in the system. Children incorrectly apply their life-world experience that, when two quantities are added together, the total is the arithmetic sum of the parts.

A number of misconceptions have been identified in recent years using assessment techniques such as clinical interviews, interviews about instances (Osborne and Freyberg, 1985), word association tests, and asking learners to write a definition of selected science terms. The results reported in the literature are usually those for which misconceptions are widespread or have an interesting distribution (often U-shaped). Some of the misconceptions which have been identified using these techniques are that (a) the earth is flat, (b) plants get food from the soil, (c) energy is a kind of fluid that moves from place to place, and (d) a bulb can be illuminated by merely connecting a wire from a battery to the bulb.

Much of this assessment work results in a set of categories that describe the variety of false views held by children. For example, Stead (1981) classified the predominant views of energy held by children as (a) pertaining to living things, (b) something that gets used up, (c) a substance-like material, and (d) something that comes from food, the body, soil, and the sun. She concluded that children find the energy concept difficult to understand.

The use of assessment to determine what students know and why has direct links to the improvement of instruction. This process (sometimes called formative evaluation) guides teaching and makes it more effective by informing teachers of the constructs which students possess. For example, it is commonly found that children's conceptions are very resistant to permanent change. Because of this situation, some have argued that science instruction should begin by assessing the life-views of children and should incorporate these life-views into instruction rather than focusing on a symbolic and abstract view of science (Duit, 1984). A gradual acceptance by students of the ways in which scientists see the world is sought.

Perhaps the most important implication for instructional improvement at present is the heightened sensitivity that this kind of assessment provides to teachers about their students. Knowing the misconceptions held by students and their operational stage of development provides powerful aids to the science teacher. What remains to be done is to make this information more readily available to teachers and to provide new ways to incorporate this knowledge into the classroom.

In a response to a variety of pressures, testing has changed in major ways during the past two decades as researchers and decision makers have worked to improve science education by improving assessment techniques. Some of these changes include:

1. *the assessment of a wider range of educational outcomes.* This change includes efforts to measure higher-order thinking skills, the use of hands-on or practical assessment procedures, understanding of the processes of science, and greater attention to affective outcomes (for example, attitudes toward science).

2. *a broader view of the purposes of testing*. Rather than being used only for evaluating student progress for promotion and admission decisions, assessment now is being used for program evaluation, monitoring the educational achievement of districts, states, and nations, and comparing the performance of various subgroups. There has been a discernible shift from student-based assessment to one that is program-based and policy relevant.

3. *the development of new techniques to probe student understanding as a basis for improving instruction*. Researchers interested in the cognitive development of children and the misconceptions which they hold have turned to other assessment techniques, particularly in-depth or clinical interviews, to assess what children believe and why. This information can be used to inform teaching and illustrate what knowledge is of most importance.

CURRICULUM EVALUATION IN SCIENCE

Background and Extent of Evaluation

The same pressures that prompted changes in testing, namely, curriculum reform and accountability, were responsible for the initiation and rapid growth of program or curriculum evaluation. The pressures were particularly acute in science and mathematics because these subjects were the targets for most of the reform efforts following the launching of Sputnik in the late 1950s. By 1977, more than 1,000 curriculum development projects in science and mathematics had been identified and listed in the *Report of the Clearinghouse on Science and Mathematics Curricular Developments* (Lockard, 1977). About half of the curriculum projects were in the United States, with an equal number located in other countries. Approximately three quarters of the projects were at the preuniversity level and two thirds of these were science projects (Welch, 1979). Thus, in the twenty years following Sputnik, about 500 preuniversity science projects had been identified by Lockard. Although most of these curriculum development projects were limited local efforts, a number of large-scale national efforts were supported by education ministries, private corporations and foundations, the National Science Foundation in the United States, and the Nuffield Foundation in Great Britain.

Evaluation activity was virtually nonexistent in the early projects. By the late 1960s and the early 1970s, however, evaluation was a component of several of the development efforts, particularly for the national reform efforts. By 1968, 19 of 68 projects listed in the *Clearinghouse* for that year (Lockard, 1968) reported that they possessed evidence of success in achieving their stated objectives. Among the exemplary curriculum evaluation efforts during these early years were the evaluations of Harvard Project Physics, the Australian Science Education Project, Scottish Integrated Science Course, Nuffield High School Biology, Science 5-13, and the Israel High School Biology Project. By 1977, nearly all large curriculum development projects in science were using evaluation in the development, implementation, and marketing of their products.

In the decade following 1977, curriculum development efforts diminished along with the attendant evaluation work. However, there has been a flurry of activity in recent years and the demand for evaluation has increased as well. For example, in the 1991 authorization bill for the National Science Foundation the U.S. Congress called for extensive evaluation of the Foundation's teacher training and curriculum development program and provided $5 million to carry out the work. In addition, other agencies that now fund curriculum development, such as private foundations, usually require an evaluation component as part of the initial proposal.

Although the *Clearinghouse* is no longer published, there are other indicators of the growth of curriculum evaluation. For example, in 1965, there were only three sessions on evaluation at the Annual Meeting of the American Educational Research Association. By 1972, the number had risen to forty and reached a peak of about eighty in the late 1970s. The number of sessions devoted to evaluation diminished somewhat during the 1980s, but today evaluation still holds steady at about forty sessions at each annual meeting of this organization.

A similar situation exists with presentations at the annual conference of the National Association for Research in Science Teaching (NARST). After experiencing a peak in the 1970s, the number of sessions on evaluation in science has leveled off currently to about 10 percent of the total. For example, in 1983, sixteen of the 118 presentations (14 percent) at the NARST meeting were on evaluation topics; in 1989, about 10 percent of the 198 papers presented were related to evaluation.

Another way in which to document the extent of evaluation and assessment activity is to examine the nature of topics that are the focus of dissertation research. This information was available for Australia and the United Kingdom. In addition, the nature of the research carried out by science educators in Israel also was studied (Tamir, 1989).

The list of science education dissertations in Australian universities published in *Studies in Science Education* (Jenkins, 1989) was examined to determine what proportion dealt with issues of assessment and/or evaluation. Of a total of 145 theses submitted between 1980 and 1986, twenty (14 percent) focused on evaluation topics. Similar findings were obtained for theses and dissertations submitted for higher degrees in British universities in the academic years 1986-1987 and 1987-1988. Among the 249 theses listed, twenty-seven theses (11 percent) examined topics in evaluation or assessment.

Tamir's (1989) survey of science education research in Israel revealed that 18 percent of the publications since 1963 were evaluation studies. Summaries of research in science education reported in the same volume for France and the Caribbean were not possible to quantify in this manner. However, both authors reported limited work in this area in their country and/or region.

Based upon the foregoing discussion, it appears that, although curriculum evaluation is not as prevalent as it once was, it still comprises a substantial proportion of the current and recent disciplined inquiry in science education.

Nature of Evaluation Efforts

It is difficult to portray all of the evaluation work in science education because of its diversity. However, several themes seem to emerge based upon the experiences of those who have reflected upon the process. Probably the most important of these is the formative-summative distinction first described by Scriven (1967). Formative evaluation, the kind which is used most often in curriculum development, involves the gathering of information for program improvement. The audience for the evaluation is the program staff and, in most cases, the evaluation is carried out by an individual or team internal to the project. Summative evaluation is the gathering of information to make decisions about the merit or worth of the final product. The audiences are potential consumers or others interested in issues of accountability such as funding agencies. Summative evaluations can be carried out by either internal or external evaluators, although an external evaluation appears to carry more credibility with the intended audiences.

Scriven's (1967) definition of the roles of evaluation includes a third component which is seldom used but which still is important for this discussion. It is "monitoring," or evaluation carried out to provide information to funding agencies.

The key distinction among the three roles is the evaluation's audience (developer, consumer, or agency). These different roles for evaluation require different information. For example, in formative evaluation, one often is interested in identifying elements of a curriculum that are weak or need revision. On the other hand, the summative evaluator seeks indicators of strength. Funding agencies, or monitoring evaluators, require sufficient justification for continued funding or termination.

Formative evaluation. Although accounts of formative evaluations usually do not meet the criteria for publication in traditional research journals, there are a few examples that illustrate the process in science curriculum development (Fraser and Cohen, 1989; Harlen, 1975; Welch, 1972). In addition, the several reports of the *Clearinghouse* describe the nature of some formative evaluation activities.

Formative evaluation is concerned with gathering information about successes and limitations of preliminary or pilot versions of curriculum materials. In theory, results from the formative phase are furnished to authors to assist them in determining necessary revisions. Some examples of the kinds of questions asked by formative evaluators are listed below.

- Is the reading level of the printed material appropriate for the target audience?
- What suggestions do teachers have for improving the teaching of the curriculum?
- Has enough material been provided to keep teachers and students fully occupied for the duration of the course?
- Does an analysis of the concepts included in the curriculum suggest a logic of presentation and organization?
- What parts of the course do the students enjoy, and for what reasons? Which parts do they dislike, and why?

• Do the results of the unit achievement tests indicate specific areas in which students are having difficulty and which, therefore, should be considered for revision?

These questions and others of a similar nature lead to formative evaluation activities that are intended to provide information that will help improve the quality and effectiveness of the curriculum materials. Some of the more common formative activities include:

1. feedback on effectiveness provided by teachers who are selected to pilot test preliminary versions of curriculum materials in their classes. The teachers selected for this process are usually highly qualified science teachers, and often become involved in the development process. Their reports can be verbal or written.
2. classroom visits by the development staff who observe children interacting with the curriculum materials.
3. interviews with students who have used the new curriculum materials.
4. teachers' and students' responses to questionnaire items dealing with various aspects of the curriculum materials such as length, comprehensibility, and readability.
5. achievement tests that measure students' performance related to the objectives (intended outcomes) of the new courses.
6. measures of other expected student outcomes resulting from undertaking the new curriculum, including attitudes toward science, understanding of science processes, critical thinking, and cognitive preferences.
7. analysis of printed materials by content specialists, layout experts, editors, and graphic designers.

Welch (1972) cautioned formative evaluators that gathering test information from students during large-scale pilot testing requires three to four months. In many instances, curriculum materials need to be revised before the test information becomes available. Furthermore, some writers of curriculum materials are suspicious of test items and of the students' inability to understand the relevant content. In these cases, writers are more likely to criticize the items than to seek to improve their own writing. And even when the writers do believe the test results, they may be unable to rewrite the appropriate text to make the concept easier to understand. Recognizing these difficulties will help make formative evaluation more useful in the improvement process. Based upon their work with the Australian Science Education Project, Fraser and Cohen (1989) presented several important recommendations for those conducting formative evaluations of science curricula:

1. A variety of evidence used in evaluation is needed to meet the needs of different audiences.
2. Evaluators need to be specific when providing feedback to writers.
3. Direct observation of classrooms by course developers provides useful and low-cost feedback.
4. Separate tests for evaluating student performance and for curriculum evaluation are desirable.
5. Results on individual test items are more useful than total test scores.
6. Both quantitative and qualitative information are useful.
7. Findings that are distilled and interpreted for authors are more useful than raw data.
8. Formative evaluators need to be cognizant of production deadlines and provide timely information.

Summative evaluation. Summative evaluation is the gathering of information to determine the value of a finished curriculum. This information could be used by potential consumers to help them in adoption decisions, by funding agencies needing to provide accountability information to their Board of Directors, and by project developers to support their marketing claims that their program is successful.

Most early summative evaluations used student achievement in science as the main indicator of success. Students in the new or experimental curriculum were given a pretest and posttest assessing mastery of the content included in the curriculum. Another group, using a traditional science curriculum, often was given the same test, so that the gains of the experimental group could be compared with the gains in the comparison group. In most instances, students studying the new curriculum did better on the new test. In hindsight, this is not surprising given that those students had been exposed to much of the content of the new test, whereas the comparison group had not.

The later curriculum evaluations expanded this notion of summative evaluation to encompass a broader range of student outcomes, including understanding the nature of the scientific enterprise, inquiry skills such as formulating hypotheses and designing experiments, and attitudes toward science and school. The evaluation of Harvard Project Physics (HPP), a secondary school physics course, illustrates this experimental approach using multiple measures of student outcomes (Welch, 1973). A total of eleven different instruments were used in this evaluation to measure student outcomes and to describe the nature of the classroom environment (see chapter 6 of this volume for a detailed discussion of the assessment, determinants, and effects of classroom environment). Tests included measures of achievement in physics, understanding of science processes, interest in science, satisfaction with the course, reaction to the course, attitudes toward physics, the learning environment, and activity in science. Measures of change were given as pretests and posttests, while those designed to describe the nature of the physics classroom were given as a midtest. A randomized data collection process was used to reduce response burden. That is, only a third of the students in a given class took each test, thus permitting three tests to be given simultaneously. Estimates of the class mean were used as the unit of analysis.

Although no differences on the cognitive measures were found between the experimental and control groups, HPP students scored higher on most of the affective measures. Relative to students undertaking traditional physics courses, HPP students were more interested in science and satisfied with the course, and they had more positive learning environments and a more positive perception of physics than did students in traditional physics courses. In addition, more than sixty research articles, dissertations, and monographs were published during the course of the evaluation of Harvard Project Physics (Welch, 1973). Examples of the kind of studies investigated included research on evaluation methodology, studies of teachers and students, and research on learning environments.

Other broad-scale summative evaluation studies carried out in connection with the development of a science curriculum include the evaluation of the Israel High School Biology Project (Tamir, 1985) and the Australian Science Education Project. As with Harvard Project Physics, these projects had an ongoing commitment to research into science teaching and learning that was woven into the primary evaluation effort. Much of the research was based on the correlates or predictors of student outcomes (for example, teacher characteristics and behaviors, learning environments, or verbal interaction in the classroom). The existence of large databases, computer facilities, contacts with schools, multiple measures of teachers and students, and evaluation staffs who were raised on a research tradition all made this merging of research and evaluation a natural development.

Curriculum research. The notion that curriculum evaluation should include a research component on the factors related to student learning has continued and in recent years has been expanded to include issues of effective development, dissemination and adoption, and implementation. It is unclear whether this curriculum analysis is categorized properly as research or evaluation. However, it is discussed here because of its prevalence in science education and the role it plays in improving science teaching and learning.

An example of this work is provided by the evaluation of the Scottish Integrated Science project (Brown, 1985). This project has been adopted by most of the secondary schools in Scotland, and adapted for use in the Caribbean, Asia, and Africa. The research focused on curriculum innovations in general rather than being concerned with a specific evaluation of the program. The success of program implementation was explored at three levels:

1. the extent to which teachers and administrators understood the various innovations and were willing to use them;
2. the extent to which the innovations were implemented in the classroom; and
3. the extent to which the implementation of the innovations led to the intended improvements in students' learning.

The work of Brown and her colleagues suggests the following strategies for effective innovation: (1) real involvement of teachers in the planning process; (2) clear understanding on the part of teachers of what changes are intended and how they are expected to achieve them; and (3) personalization of the intended innovation by the teachers (that is, the teachers must perceive the innovation as salient to their own aspirations and problems).

Studies of adoption and implementation have been carried out in connection with the Australian Science Education Project (Owen, 1979) and the Nuffield projects in the United Kingdom (Tall, 1981). Implementation of new curricula has been difficult to achieve and maintain. The demands of day-to-day operation of schools, the lack of support for teachers, and the apparent reluctance of publishing companies to risk change are examples of factors that support maintenance of the status quo.

Recognizing many of these barriers, several new curriculum development efforts have been supported by the National Science Foundation in the United States and require a partnership among school districts, publishing companies, and educational institutions such as universities or curriculum development centers. These so-called "triad" projects were evaluated extensively during the development and early implementation process to provide information to the NSF to inform their future development programs.

Another kind of research carried out in connection with a curriculum development project is illustrated by the Learning in Science Project in New Zealand (Osborne and Freyberg, 1985). The research identified the alternative frameworks held by children and used this information to produce teaching and learning activities to help students construct understanding about the natural world which was meaningful and useful to them. Curriculum development was based heavily on research that examined the mismatch between the logical structure of science and the way in which people learn.

Curriculum evaluation has occupied a prominent position in science education during the past two decades. A variety of formative and summative evaluation activities have been implemented to improve materials and assist adoption/adaption decisions. In addition, research has been carried out as part of the evaluation effort to improve the teaching and learning process and to improve our understanding of effective development, implementation, and utilization.

CONCLUSIONS AND RECOMMENDATIONS

In this chapter I have examined developments in student assessment and curriculum evaluation in preuniversity science education. Several general findings have emerged:

1. Considerable activity during the past two decades has generated a large body of literature on assessment and evaluation. Much of this activity was carried out in connection with large curriculum development projects.

2. Improved techniques for student assessment and curriculum evaluation have been developed and implemented.

3. Many of these improvements have been associated with large and sustained evaluation efforts for which a team of specialists worked on the problems over a period of several years.

4. Support for assessment and evaluation has been provided by local, state, federal, and private agencies in response to public demands for improved education and accountability.

5. Student assessment and curriculum evaluation gradually have become more responsive to audience needs and concerns as ways have been sought to increase the utilization of results.

In addition, several specific assessment and evaluation trends have been noted. Student assessment has expanded from a heavy emphasis on student achievement to include attitudes, understanding of science processes, practical skills, and history and philosophy of science, among others. Inclusion of such topics in assessment instruments sensitizes students and teachers to the importance of these concepts.

Furthermore, the role of testing has shifted from a rather narrow focus on standardized tests for decisions on promotion and admission to the use of assessment procedures for program review and evaluation; state, national, and international assessments; educational accountability; and the improvement of instruction. Changes have occurred in both the nature and purpose of science assessment.

New techniques for assessment have been developed and implemented for in-depth probing of what students know and how they know it. Concerns with cognitive development, student misconceptions and their role in science instruction, and the role of prior knowledge in learning all have required the development and use of different forms of investigation, particularly the in-depth or clinical interview. Greater understanding of teaching and learning has resulted.

Advances in curriculum evaluation have occurred as well. Thirty years ago, the words "formative" and "summative" evaluation did not exist. Today, they are major concepts in the field of evaluation. Substantial effort has been exerted in using evaluation to improve curriculum quality (formative) and providing useful information to consumers facing adoption/adaption decisions (summative). In addition, this evaluation has generated data which have been analyzed to yield new knowledge about the teaching/learning process and the development, adoption, implementation, and persistence of innovations, and it has led to improvements in the techniques of assessment and evaluation.

Emerging from the discussion in this chapter are the following recommendations for improving student assessment and curriculum evaluation in science education:

1. The traditional emphasis on the evaluation of student learning should be supplemented by greater attention to formative curriculum evaluation (aimed at improving the quality of instruction) and summative curriculum evaluation (aimed at guiding adoption and adaption decisions).

2. Evaluation efforts should extend beyond a focus on student achievement to encompass other valued science outcomes such as attitudes, inquiry skills, problem solving, and understanding of the nature of science.

3. Traditional paper-and-pencil evaluation instruments should be complemented by the use of alternative and authentic evaluation techniques such as in-depth interviews, portfolio assessments, and practical performance tests.

4. In order to improve the usefulness of formative curriculum evaluation, a variety of qualitative and quantitative evaluation methods should be used; highly specific information that can guide improvements should be sought; separate student and curriculum evaluation instruments should be used; and performance on individual test items should be considered as well as scores on the total test.

There are still important issues to be addressed in the future. First, the impact of assessment and evaluation needs to be examined in more detail. In a review of the literature on the unintended effects of program evaluation (Welch and Sternhagen, 1991), virtually no information was found about the impact of evaluation. Considerable concern has been directed to the impact of testing, but the emphasis has been on standardized testing rather than on the kind of assessment described here.

Second, further improvement is needed in the assessment and evaluation process. More attention is needed to addressing audience concerns, clearer statements of purpose, a better understanding of the role of evaluation in policymaking, and the development of more effective methods.

Third, more work is needed on using assessment to improve instruction. Improvements are needed in practical examinations, authentic or performance-based assessment (that is, testing science in the spirit of science), and measurement of student attitudes, as well as in the ways in which assessment findings can be linked effectively in the day-to-day operation of the classroom. This is true for large-scale assessments and for the kinds of clinical interviews being used for determining students' conceptions.

Finally, evaluations must have greater impact on curriculum developers and users. Information must be provided to decision makers in a timely manner and in a form that is easily used. To borrow a term from the computer field, assessment and evaluation have to become more "user friendly." Achieving this goal will make assessment of students and evaluation of curricula even more useful for the improvement of science education.

REFERENCES

Black, Paul J. "Performance Assessment and Accountability: The Experience in England and Wales." Paper presented at the Annual Meeting of the American Educational Research Association, Atlanta, 1993.

Brown, Sally. "A Research Approach to the Evaluation of Scottish Integrated Science." In *The Role of Evaluators in Curriculum Development*, edited by Pinchas Tamir, pp. 122-141. London, England: Croom Helm, 1985.

Buros, Oscar K., ed. *Eighth Mental Measurements Yearbook: Volume II*. Highland Park, N.J.: Gryphon Press, 1978.

Darling-Hammond, Linda, and Lieberman, Ann. "The Shortcomings of Standardized Tests," *Chronicle of Higher Education*, 29 January 1992, pp. 8-9.

Duit, Reinders. "Learning the Energy Concept in School—Empirical Results from the Philippines and West Germany," *Physics Education* 19 (1984): 59-66.

Fraser, Barry J., and Cohen, David. "A Retrospective Account of the Development and Evaluation Processes of a Science Curriculum Project," *Science Education* 73 (1989): 25-44.

Harlen, Wynne. "A Critical Look at the Classical Strategy Applied to Formative Evaluation," *Studies in Educational Evaluation* 1 (1975): 37-53.

Hueftle, Stacey J.; Rakow, Steven J.; and Welch, Wayne W. *Images of Science: A Summary of Results from the 1981-82 National Assessment in Science*. Minneapolis: College of Education, University of Minnesota, 1983.

IAEP (International Assessment of Educational Progress). *A World of Differences: An International Assessment of Mathematics and Science.* Princeton, N.J.: Educational Testing Service, 1989.

Jenkins, Edgar W., ed. "Research Notes," *Studies in Science Education* 16 (1989): 116-193.

Kulm, Gerald, and Malcom, Shirley, eds. *Science Assessment in the Service of Reform.* Washington, D.C.: American Association for the Advancement of Science, 1991.

Lockard, J. David, ed. *Sixth Report of the International Clearinghouse on Science and Mathematics Curricular Developments 1968.* College Park: University of Maryland, 1968.

Lockard, J. David, ed. *Twenty Years of Science and Mathematics Curriculum Development. The 10th Report of the International Clearinghouse.* College Park: University of Maryland, 1977.

NAEP (National Assessment of Educational Progress) *Learning by Doing.* Report No. 17-HOS-80. Princeton, N.J.: Educational Testing Service, 1987.

Osborne, Roger, and Freyberg, Peter. *Learning Science: The Implications of Children's Science.* Auckland, New Zealand: Heinemann, 1985.

Owen, John M. *The Impact of the Australian Science Education Project on Schools.* Canberra: Curriculum Development Centre, 1979.

Robitaille, David F.; Schmidt, William H.; Raizen, Senta; McKnight, Curtin; Britton, Edward; and Nicol, Cynthia. *Curriculum Frameworks for Mathematics and Science.* Third International Mathematics and Science Study, Monograph No. 1. Vancouver, B.C., Canada: Pacific Educational Press, 1993.

Scriven, Michael. "The Methodology of Evaluation." In *Perspectives of Curriculum Evaluation,* edited by Robert E. Stake, pp. 39-83. Chicago: Rand McNally, 1967.

Stead, Beverley. "Ecology, Energy, and the Forms 1-4 Science Syllabus," *New Zealand Science Teacher* 28 (1981): 17-20.

Strauss, Sidney, and Stavy, Ruth. "Educational-Developmental Psychology and Curriculum Development: The Case of Heat and Temperature." In *Proceedings of the International Seminar on Misconceptions in Science and Mathematics,* edited by Hugh Helm and Joseph D. Novak, pp. 310-321. Ithaca, N.Y.: Cornell University, 1983.

Tall, Gene E. "British Science Curriculum Projects: How Have They Taken Root in the Schools?" *European Journal of Science Education* 3 (1981): 17-38.

Tamir, Pinchas. "The Evaluation of the Israel High School Biology Project." In *The Role of Evaluators in Curriculum Development,* edited by Pinchas Tamir, pp. 162-183. London, England: Croom Helm, 1985.

Tamir, Pinchas. "Research in Science Education in Israel," *Studies in Science Education* 16 (1989): 229-238.

Tamir, Pinchas, et al. "Testing and the School Curriculum: Evolving Trends," *Studies in Educational Evaluation* 13 (1987): 3-103.

Welch, Wayne W. "Some Problems in Evaluating a National Curriculum Project," *Curriculum Theory Network* 1 (1972): 232-241.

Welch, Wayne W. "Review of the Research and Evaluation Program of Harvard Project Physics," *Journal of Research in Science Teaching* 10 (1973): 365-378.

Welch, Wayne W. "Twenty Years of Science Curriculum Development: A Look Back." In *Review of Research in Education*, vol. 7, edited by David Berliner, pp. 282-306. Washington, D.C.: American Educational Research Association, 1979.

Welch, Wayne W. *Blueprint for Reform: Assessment*. Washington, D.C.: Project 2061, American Association for the Advancement of Science, in press.

Welch, Wayne W., and Sternhagen, Fred. "Unintended Effects of Program Evaluation," *Evaluation Practice* 12, no. 2 (1991): 121-129.

White, Richard P., and Tisher, Richard T. "Research on Natural Sciences." In *Handbook of Research on Teaching*, 3d ed., edited by Merlin C. Wittrock, pp. 874-905. New York: Macmillan, 1986.

Chapter 6

CLASSROOM LEARNING ENVIRONMENTS

Barry J. Fraser and Theo Wubbels

Science educators often speak of classroom environment (also referred to as climate, atmosphere, tone, ethos, or ambience) and they consider it to be important in its own right and influential in terms of students' learning. Despite the fact that classroom environment is a somewhat subtle concept, much progress has been made over the last quarter century in conceptualizing it, assessing it, and researching its determinants and effects. Researchers in science education have made many important contributions to the field by developing, validating, and applying instruments for assessing the environment in classrooms.

Many questions of interest to teachers, educational researchers, curriculum developers, and policymakers in science education can be asked about classroom environments. Does a classroom's environment affect student learning and attitudes? What is the impact of a new curriculum or teaching method on classroom environment? Can teachers conveniently assess the climate of their own classrooms and can they change this environment? What are some of the determinants of classroom environment? Is there a discrepancy between students' perceptions of the actual and their preferred classroom environment? If so, does this discrepancy matter in terms of student outcomes? Do teachers and their students perceive the same classroom environment similarly? These questions represent the thrust of the work on educational environments over the past thirty years and constitute the main areas considered in this chapter.

Chapter consultants: Peter Okebukola (Lagos State University, Nigeria) and Frances Lawrenz (University of Minnesota, USA).

Traditionally research and evaluation in science education have tended to rely heavily and sometimes exclusively on the assessment of academic achievement and other valued learning outcomes. Although few educators would dispute the worth of outcome measures, such measures cannot give a complete picture of the educational process. In this chapter, we concentrate on one approach to conceptualizing, assessing, and investigating what happens to students during their schooling. The main focus is upon students' and teachers' perceptions of important social and psychological aspects of the learning environments of science classrooms and the implications of this research for improving science education.

In the introductory section of this chapter we provide background information about the field of classroom environment. In subsequent sections we consider various instruments used to assess perceptions of psychosocial environment, provide an overview of several lines of past research in this field, and show how teachers can use instruments that assess classroom environment in their attempts to improve their own classrooms. After noting recent developments in research on classroom environment, we draw some implications of that research for improving science education.

CLASSROOM ENVIRONMENT AS A FIELD OF STUDY

Approaches to Studying Learning Environments

Three common approaches to studying classroom environment involve systematic observation, case studies, and assessments of students' and teachers' perceptions. Perceptual measures form the major focus in this chapter. This approach has several merits (Fraser, 1986). First, paper-and-pencil perceptual measures are more economical than classroom observation techniques that involve the expense of trained outside observers. Second, perceptual measures are based on students' experiences in many classes, while observational data usually are restricted to a very small number of classes. Third, perceptual measures involve the pooled judgments of all students in a class, whereas observation techniques typically involve only a single observer. Fourth, because students' perceptions are important determinants of students' behavior, they are more valuable in explaining that behavior than are inferences about behavior made by external observers. Fifth, perceptual measures of classroom environment typically have been found to account for considerably more of the variation in students' learning than have directly observed variables.

Historical Background

It is now over a quarter of a century since the *Learning Environment Inventory* was used as part of the research and evaluation activities of Harvard Project Physics (Welch and Walberg, 1972). About the same time, Moos began developing social climate scales for a wide variety of human environments, including the *Classroom Environment Scale* for use in school settings (Moos, 1974). The way in which these two programs of research have developed and spawned many new lines of research on learning environments is reflected in several comprehensive reviews of the literature (Fraser, 1986, 1994; Fraser and Walberg, 1991).

While we focus here upon studies of classroom environment done in the past twenty-five years or so, we fully acknowledge that this work builds upon and has been influenced by earlier work in two areas. First, we recognize the influence of the momentous theoretical, conceptual, and measurement foundations laid half a century ago by pioneers like Kurt Lewin and Henry Murray and their followers, such as C. Robert Pace and George Stern (see Fraser, 1986). Second, research involving assessments of perceptions of classroom environment epitomized in the work described in this chapter also was influenced by prior work involving low-inference, direct-observational methods for measuring classroom climate.

Distinction Between School and Classroom Environment

It is useful to distinguish classroom-level environment from school-level environment. The latter involves psychosocial aspects of the climate of whole schools (Fraser, Williamson, and Tobin, 1987). Despite their simultaneous development and logical linkages, the fields of classroom-level and school-level environment have remained remarkably independent. Although the focus of this chapter is restricted primarily to classroom environment, we acknowledge that it would be desirable to break away from the existing tradition of independence and to achieve a better integration of the fields of school and classroom environment.

School climate research owes much in theory, instrumentation, and methodology to earlier work on organizational climate in business contexts. For example, two widely used instruments in school environment research (Halpin and Croft's [1963] *Organizational Climate Description Questionnaire* [OCDQ] and Stern's [1970] *College Characteristic Index* [CCI]) relied heavily on previous work in business organizations. Consequently, one feature of school-level environment work which distinguishes it from classroom-level environment research is that the former has tended to be associated with the field of educational administration and to rest on the assumption that schools can be viewed as formal organizations. Another distinguishing feature is that, whereas classroom-level research has focused on secondary and elementary schools rather than higher education, a sizeable proportion of school-level environment research has involved the climate of higher education institutions.

Some promising recent work has combined the use of classroom and school environment measures to advantage within one study (Fraser, Williamson, and Tobin, 1987). Another study has involved the use of school climate scales to reveal interesting differences between elementary and secondary schools (Docker, Fraser, and Fisher, 1989). Other research has involved the successful application of the methods of improving classroom-level environments described in this chapter to the improvement of school-level environments (Fraser, 1994). Overall, this recent research attests to the value of school climate research and suggests that the time is ripe for a better integration of the research on classroom environment with that on school environment. An understanding of the links among the environments of the class and school, together with linkages between these settings and other important environments such as the home, could lead to improvements in education generally, including science education.

Measurement Level

Characteristics of the learning environment can be measured through the perceptions of students, teachers, or observers to produce distinct variables each of which has its own significance. Assessment involving students' perceptions can be subdivided further into whether it involves the individual student's perceptions or the intersubjective perceptions of all students in the same class. This distinction in past classroom environment research often has been important when choosing an appropriate unit of statistical analysis such as individual student scores or class mean scores. (See Fraser, 1986.) Because of the great advances that have been made recently in *multilevel analysis* (Goldstein, 1987), more sophisticated techniques are now available for analyzing the typical data (for example, on students nested within classes) found in much research on learning environments.

INSTRUMENTS FOR ASSESSING CLASSROOM ENVIRONMENT

Among the instruments most frequently used in prior research to assess perceptions of classroom learning environment are the following:

Learning Environment Inventory (LEI)
Classroom Environment Scale (CES)
Individualized Classroom Environment Questionnaire (ICEQ)
My Class Inventory (MCI)
College and University Classroom Environment Inventory (CUCEI)
Science Laboratory Environment Inventory (SLEI)
Questionnaire on Teacher Interaction (QTI)

Each of the above instruments is suitable for convenient group administration and can be scored either by hand or by computer.

Table 6-1 shows the name of each scale contained in each instrument, the level (elementary, secondary, higher education) for which each instrument is suited, the number of items contained in each scale, and the classification of each scale according to Moos's (1974) scheme for classifying human environments. Moos's three basic types of dimension are "Relationship Dimensions" (which identify the nature and intensity of personal relationships within the environment and assess the extent to which people are involved in the environment and support and help each other), "Personal Development Dimensions" (which assess basic directions along which personal growth and self-enhancement tend to occur), and "System Maintenance and System Change Dimensions" (which involve the extent to which the environment is orderly, clear in expectations, under control, and responsive to change).

Table 6-1

Overview of Scales Contained in Seven Classroom
Environment Instruments
(LEI, CES, ICEQ, MCI, CUCEI, SLEI, AND QTI)

			Scales Classified According to Moos's Scheme		
Instrument	Level	Items Per Scale	Relationship Dimensions	Personal Development Dimensions	System Maintenance & Change Dimensions
Learning Environment Inventory (LEI)	Secondary	7	Cohesiveness Friction Favoritism Cliqueness Satisfaction Apathy	Speed Difficulty Competitiveness	Diversity Formality Material Environment Goal Direction Disorganization Democracy
Classroom Environment Scale (CES)	Secondary	10	Involvement Affiliation Teacher Support	Task Orientation Competition	Order & Organization Rule Clarity Teacher Control Innovation
Individualised Classroom Environment Questionnaire (ICEQ)	Secondary	10	Personalization Participation	Independence Investigation	Differentiation
My Class Inventory (MCI)	Elementary	6-9	Cohesiveness Friction Satisfaction	Difficulty Competitiveness	
College and University Classroom Environment Inventory (CUCEI)	Higher Education	7	Personalization Involvement Student Cohesiveness Satisfaction	Task Orientation	Innovation Individualization
Science Laboratory Environment Inventory (SLEI)	Upper Secondary Higher Education	7	Student Cohesiveness	Open-Endedness Integration	Rule Clarity Material Environment
Questionnaire on Teacher Interaction (QTI)	Secondary Elementary	8-10	Helpful/Friendly Understanding Dissatisfied Admonishing		Leadership Student Responsibility & Freedom Uncertain Strict

Learning Environment Inventory (LEI). The initial development and validation of a preliminary version of the LEI began in the late 1960s in conjunction with the evaluation of and research on Harvard Project Physics (Fraser, Anderson, and Walberg, 1982). In selecting the fifteen climate dimensions, the developers included as scales only concepts previously identified as good predictors of learning, concepts considered relevant to social psychological theory and research, concepts similar to those found useful in theory and research in education, or concepts intuitively judged relevant to the social psychology of the classroom. The final version of the LEI contains a total of 105 statements (seven per scale) descriptive of typical school classes. The respondent expresses degree of agreement or disagreement with each statement on a four-point scale with response alternatives of Strongly Disagree, Disagree, Agree, and Strongly Agree. The scoring direction (or polarity) is reversed for some items. A typical item contained in the Cohesiveness scale is: "All students know each other very well." An item from the Speed scale is: "The pace of the class is rushed."

Classroom Environment Scale (CES). The CES was developed by Rudolf Moos at Stanford University (Moos and Trickett, 1987) and grew out of a comprehensive program of research involving perceptual measures of a variety of human environments including psychiatric hospitals, prisons, university residences, and work milieus (Moos, 1974). Moos and Trickett's (1987) final published version of the CES contains nine scales with ten items of True-False response format in each scale. Published materials include a test manual, a questionnaire, an answer sheet, and a transparent key for hand scoring. Typical items in the CES are: "The teacher takes a personal interest in the students" (Teacher Support) and "There is a clear set of rules for students to follow" (Rule Clarity).

Individualized Classroom Environment Questionnaire (ICEQ). The ICEQ differs from other classroom environment scales in that it assesses those dimensions (for example, Personalization, Participation) that distinguish between individualized classrooms and more conventional classrooms. The initial development of the long form ICEQ (Fraser, 1990) involved (a) choosing dimensions that characterize the classroom learning environment as it is described in the literature of individualized and open education, (b) extensive interviewing of teachers and secondary school students to ensure that the ICEQ dimensions and individual items were considered salient, (c) modifying items after receiving reactions from selected experts, teachers, and junior high school students, and (d) field testing the instrument and conducting item analyses in order to identify items whose removal would improve scale statistics (for example, reliability). The final published version of the ICEQ (Fraser, 1990) contains fifty items altogether, with ten items for each of the five scales. Each item is responded to on a five-point scale with the alternatives of Almost Never, Seldom, Sometimes, Often, and Very Often. The scoring direction is reversed for many of the items. Typical items are: "The teacher considers students' feelings" (Personalization) and "Different students use different books, equipment, and materials" (Differentiation). The published form of the ICEQ consists of a handbook and test master sets from which unlimited numbers of copies of the questionnaires and response sheets may be made.

My Class Inventory (MCI). The LEI has been simplified to form the MCI which is suitable for children 8 to 12 years old (Fraser, Anderson, and Walberg, 1982; Fraser and O'Brien, 1985). Although the MCI was developed originally for use at the elementary school level, it also has been found to be very useful with students in the junior high school, especially those who might experience reading difficulties with the LEI. The MCI differs from the LEI in four important ways. First, in order to minimize fatigue among younger children, the MCI contains only five of the fifteen scales in the LEI. Second, item wording has been simplified to enhance readability. Third, the four-point response format in the LEI has been reduced to a two-point (Yes-No) response format. Fourth, students answer on the questionnaire itself instead of on a separate response sheet to avoid errors in transferring responses from one place to another. The final form of the MCI contains thirty-eight items. Typical items are: "Children are always fighting with each other" (Friction) and "Children seem to like the class" (Satisfaction). The reading level of these MCI items is well suited to students at the elementary school level.

College and University Classroom Environment Inventory (CUCEI). Although some notable prior work has focused on the institutional-level or school-level environment in colleges and universities (for example,

Halpin and Croft, 1963; Stern, 1970), surprisingly little work has been done in higher education classrooms that is parallel to the traditions of classroom environment research at the secondary and elementary school levels. Since one likely explanation for this shortage is simply the unavailability of a suitable instrument, the CUCEI was developed to fill this void. It is to be used in small classes, but not in lectures or laboratory classes (Fraser and Treagust, 1986). The final form of the CUCEI contains seven seven-item scales. Each item has four responses (Strongly Agree, Agree, Disagree, Strongly Disagree) and polarity is reversed for approximately half of the items. Typical items are: "Activities in this class are clearly and carefully planned" (Task Orientation) and "Teaching approaches allow students to proceed at their own pace" (Individualization).

Science Laboratory Environment Inventory (SLEI). Because of the critical importance and uniqueness of laboratory settings in science education, a new instrument specifically suited to assessing the environment of science laboratory classes at the senior high school or higher education levels was developed (Fraser, McRobbie, and Giddings, in press). This new questionnaire, the SLEI, has five scales and the response alternatives for each item are Almost Never, Seldom, Sometimes, Often, and Very Often. Typical items include: "We know the results that we are supposed to get before we commence a laboratory activity" (Open-Endedness), and "The laboratory work is unrelated to the topics that we are studying in our science classes" (Integration). The Open-Endedness scale was included because of the importance of open-ended laboratory activities claimed in the literature. The SLEI was field tested simultaneously in six countries (USA, Canada, England, Israel, Australia, and Nigeria) with a sample of 5,477 students in over 269 classes in order to furnish comprehensive information about the instrument's cross-national validity and usefulness.

In order to provide the reader with a concrete example of a classroom environment instrument that is highly relevant to science education, figure 6-1 contains a complete copy of the SLEI "actual" form. (See pp. 127-128 for a clarification of the distinction between "actual" and "preferred" forms.) Responses to items in figure 6-1 with their item numbers *not* underlined are scored 1, 2, 3, 4, and 5, respectively, for the responses Almost Never, Seldom, Sometimes, Often, and Very Often. Responses to items for which the item number is underlined are scored

Remember that you are describing your **actual** classroom.	Almost Never / Seldom / Sometimes / Often / Very Often	For Teacher's Use
1. I get on well with students in this laboratory class.	1 2 3 4 5	____
2. There is opportunity for me to pursue my own science interests in this laboratory class.	1 2 3 4 5	____
3. What I do in our regular science class is unrelated to my laboratory work.	1 2 3 4 5	R ____
4. My laboratory class has clear rules to guide my activities.	1 2 3 4 5	____
5. I find that the laboratory is crowded when I am doing experiments.	1 2 3 4 5	R ____
6. I have little chance to get to know other students in this laboratory class.	1 2 3 4 5	R ____
7. In this laboratory class, I am required to design my own experiments to solve a given problem.	1 2 3 4 5	____
8. The laboratory work is unrelated to the topics that I am studying in my science class.	1 2 3 4 5	R ____
9. My laboratory class is rather informal and few rules are imposed on me.	1 2 3 4 5	R ____
10. The equipment and materials that I need for laboratory activities are readily available.	1 2 3 4 5	____
11. Members of this laboratory class help me.	1 2 3 4 5	____
12. In my laboratory sessions, other students collect different data than I do for the same problem.	1 2 3 4 5	____
13. My regular science class work is integrated with laboratory activities.	1 2 3 4 5	____
14. I am required to follow certain rules in the laboratory.	1 2 3 4 5	____
15. I am ashamed of the appearance of this laboratory.	1 2 3 4 5	R ____
16. I get to know students in this laboratory class well.	1 2 3 4 5	____
17. I am allowed to go beyond the regular laboratory exercise and do some experimenting of my own.	1 2 3 4 5	____
18. I use the theory from my regular science class sessions during laboratory activities.	1 2 3 4 5	____
19. There is a recognized way for me to do things safely in this laboratory.	1 2 3 4 5	____
20. The laboratory equipment which I use is in poor working order.	1 2 3 4 5	R ____
21. I am able to depend on other students for help during laboratory classes.	1 2 3 4 5	____
22. In my laboratory sessions, I do different experiments than some of the other students.	1 2 3 4 5	____
23. The topics covered in regular science class work are quite different from topics with which I deal in laboratory sessions.	1 2 3 4 5	R ____
24. There are few fixed rules for me to follow in laboratory sessions.	1 2 3 4 5	R ____
25. I find that the laboratory is hot and stuffy.	1 2 3 4 5	R ____
26. It takes me a long time to get to know everybody by his/her first name in this laboratory class.	1 2 3 4 5	R ____
27. In my laboratory sessions, the teacher decides the best way for me to carry out the laboratory experiments.	1 2 3 4 5	R ____
28. What I do in laboratory sessions helps me to understand the theory covered in regular science classes.	1 2 3 4 5	____
29. The teacher outlines safety precautions to me before my laboratory sessions commence.	1 2 3 4 5	____
30. The laboratory is an attractive place for me to work in.	1 2 3 4 5	____
31. I work cooperatively in laboratory sessions.	1 2 3 4 5	____
32. I decide the best way to proceed during laboratory experiments.	1 2 3 4 5	____
33. My laboratory work and regular science class work are unrelated.	1 2 3 4 5	R ____
34. My laboratory class is run under clearer rules than my other classes.	1 2 3 4 5	____
35. My laboratory has enough room for individual or group work.	1 2 3 4 5	____

Figure 6-1

**Items in the Actual Form of the Science Laboratory
Environment Inventory (SLEI)**

in the reverse manner. Omitted or invalidly answered items are scored 3
(or, alternatively, assigned the mean of the other items in the same

scale). In order to make the SLEI easy to score by hand, items have been arranged in cyclic order and in blocks. That is, the total score on the Student Cohesiveness scale is obtained simply by adding the item scores for the first item in each block (items 1, 6, 11, 16, 21, 26, and 31). Similarly, the Material Environment total score is the sum of the scores for the last item in each block (items 5, 10, 15, 20, 25, 30, and 35).

Questionnaire on Teacher Interaction (QTI). In research originating in the Netherlands a learning environment questionnaire was developed to enable teacher educators to give preservice and in-service teachers advice about the nature and quality of the interaction between teachers and students (Wubbels, Brekelmans, and Hooymayers, 1991, 1992). Drawing upon Timothy Leary's theoretical model of proximity (Cooperation-Opposition) and influence (Dominance-Submission), the *Questionnaire on Teacher Interaction (QTI)* was developed to assess student perceptions of the eight behavior aspects of Leadership, Helpful/Friendly, Understanding, Student Responsibility and Freedom, Uncertain, Dissatisfied, Admonishing, and Strict behavior.

The original version of the QTI has seventy-seven items altogether (approximately ten per scale), although a more economical forty-eight-item version also exists now. Each item is responded to on a five-point scale ranging from Never to Always. Typical items are "She/he gives us a lot of free time" (Student Responsibility and Freedom behavior) and "She/he gets angry" (Admonishing behavior). The QTI has been found to be valid and reliable in studies among secondary school students in the Netherlands and the USA (Wubbels and Levy, 1991) and in Australia. Recently a version for the elementary school has been used in research in Singapore.

Preferred Forms of Scales. A distinctive feature of most of the instruments in Table 6-1 is that, in addition to a form that measures perceptions of *actual* or *experienced* classroom environment, another form measures perceptions of *preferred* classroom environment. The preferred (or ideal) forms are concerned with goals and value orientations and measure perceptions of the classroom environment ideally liked or preferred. Although item wording is identical or similar for actual and preferred forms, the instructions for answering each are different. Having different actual and preferred forms has enabled these instruments to be used for the range of new research applications discussed later in this chapter. Although the LEI and MCI originally were designed only to measure actual environment, Fraser and O'Brien (1985) have used a preferred form of the MCI successfully with elementary school classes.

Each of the instruments described above has been field tested extensively and found to be valid and reliable for applications involving either the individual student's score or the class mean score as the unit of analysis (Fraser, 1994).

TYPES OF RESEARCH USING CLASSROOM ENVIRONMENT INSTRUMENTS

In order to illustrate the range of possible uses of classroom environment scales, we consider here research involving (a) associations between student outcomes and classroom environment, (b) use of classroom environment dimensions as criterion variables (including curriculum evaluation studies and investigations of differences between students' and teachers' perceptions of the same classrooms), and (c) investigations of whether students achieve better in classes that approximate their preferred environments.

Associations between Student Outcomes and Classroom Environment

The strongest tradition in past classroom environment research has involved investigation of associations between students' cognitive and affective learning outcomes and their perceptions of their classroom environments (Haertel, Walberg, and Haertel, 1981). Numerous research programs have shown that students' perceptions account for appreciable amounts of variance in learning outcomes, often beyond that attributable to students' background characteristics. The practical implication from this research is that students' outcomes might be improved by creating classroom environments found empirically to be conducive to learning.

Fraser (1994) has tabulated a set of forty studies in which the effects of classroom environment on students' outcomes in science were investigated. This tabulation shows that studies of associations between outcome measures and perceptions of classroom environment have involved a variety of cognitive and affective outcome measures, a variety of classroom environment instruments, and a variety of samples (ranging across numerous countries and grade levels).

The findings from prior research are highlighted in the results of a meta-analysis involving twelve studies encompassing 17,805 students (Haertel, Walberg, and Haertel, 1981). Learning environment was found to be consistently and strongly associated with cognitive and affective learning outcomes, although correlations generally were higher in samples of older students and in studies employing collectivities such as classes and schools (in contrast to individual students) as the units of statistical analysis. In particular, better achievement on a variety of outcome measures was found consistently in classes perceived as having greater Cohesiveness, Satisfaction, and Goal Direction, and less Disorganization and Friction.

Fraser and Fisher (1982) reported a study of the effects of classroom environment on student outcomes involving a representative sample of 116 Grade 8 and 9 science classes, each with a different teacher, in thirty-three different schools. Three cognitive and six affective measures were administered both at the beginning and end of the same school year, while classroom environment was assessed by administering the CES and ICEQ at mid-year. In addition, information was gathered about students' general ability. Overall, the study yielded consistent support for the existence of outcome-environment relationships and suggested some important tentative implications for educators wishing to enhance students' achievement of particular outcomes by creating classroom environments found empirically to be conducive to achievement. For example, practitioners are likely to find useful the finding that Order and Organization seemed to have a positive influence on student achievement of a variety of aims.

In the research of Fraser, McRobbie, and Giddings (in press), which involved the use of the SLEI in science laboratory environments, the most striking finding was that both cognitive and affective outcomes were superior in classes with higher scores on the Integration scale (that is, the link between the work covered in laboratory classes and theory classes) than in classes with lower scores on Integration.

It is interesting to note that the research on outcome-environment relationships in developed countries has been replicated in numerous developing countries (Fraser, 1993). However, because much of this research in both developed and developing countries has been correlational, other types of studies are needed to establish causal links between classroom climate and student outcomes.

Use of Classroom Environment Perceptions as Criterion Variables

Fraser (1986) tabulated thirty-nine studies in which classroom environment dimensions were employed as dependent variables in science education in (a) curriculum evaluation studies, (b) investigations of differences between student and teacher perceptions of actual and preferred environment, and (c) studies involving other independent variables.

One promising but largely neglected use of classroom environment instruments is as a source of process criteria in evaluating innovations and new curricula (Fraser, Williamson, and Tobin, 1987). For example, a study involving an evaluation of the Australian Science Education Project (ASEP) revealed that, in comparison with a control group, students in ASEP classes perceived their classrooms as being more satisfying and individualized and as having a better material environment (Fraser, 1986). The significance of the ASEP evaluation, as well as Welch and Walberg's (1972) evaluation of Harvard Project Physics, is that classroom environment variables differentiated revealingly between curricula, even when various outcome measures showed negligible differences.

The fact that some classroom environment instruments have different actual and preferred forms that can be used either with teachers or students permits investigation of differences between students and teachers in their perceptions of the same actual classroom environment and of differences between the perceived actual environment and that preferred by students or teachers. Research into differences between forms reported by Fisher and Fraser (1983) revealed that students preferred a more positive classroom environment than was perceived to be present, and that teachers perceived a more positive classroom environment than did their students in the same classrooms. These interesting results replicate patterns in other studies in school classrooms in USA, Israel, and Australia (see Fraser, 1994), as well as in other settings such as hospital wards and work milieus (see Moos, 1974). These studies inform educators that students and teachers are likely to differ in the way in which they perceive the actual environment of the same classrooms, and that the environment preferred by students commonly differs from that they perceive to be present in classrooms.

In a study involving the use of the QTI among teachers in the Netherlands, Wubbels, Brekelmans, and Hooymayers (1992) also found that a high proportion of teachers (70 percent) viewed the learning environment more favorably than their students. However, the perceptions of the other teachers in this group (30 percent) were more negative than those of their students. Overall, this study suggested that teacher perceptions of the learning environment are shaped partly by their ideals about the learning environment (that is, teachers' ideals can distort their perceptions of the actual learning environment). The mismatch between a teacher's and his or her students' perceptions on average was larger for teachers who showed less behavior that promoted student cognitive and affective outcomes than for teachers who showed many such behaviors.

Therefore, discussions between university staff and preservice teachers about their teaching practice should not be based merely on teachers' perceptions; feedback data gathered through classroom observations or students' perceptions also are desirable. Similarly, because teachers' perceptions can depend upon teachers' beliefs, it is important that evaluations of teacher preparation and staff development programs do not rely heavily on teacher self-reports.

The third group of studies reviewed by Fraser (1986) shows that other researchers have used classroom environment dimensions as criterion variables in studies aimed at identifying how the classroom environment varies with such factors as teacher personality, class size, grade level, subject matter, the nature of the school-level environment, and the type of school.

Do Students Achieve Better in Classrooms Similar to Their Preferred Environment?

Whereas much past research has concentrated on investigations of associations between student outcomes and the nature of the actual environment, the use of both actual and preferred forms of classroom environment instruments permits exploration of whether students achieve better when there is a high similarity between the student-perceived actual classroom environment and that preferred by students. Fraser and Fisher (1983) used a person-environment interaction framework in exploring whether or not student outcomes depend not only on the nature of the classroom environment, but also on the match between students' preferences and the environment. The basic design of their study involved the prediction of posttest achievement from pretest performance, general ability, the five actual individualization variables measured by the ICEQ and five variables indicating actual-preferred interaction. The class mean was used as the unit of analysis.

Overall, the findings from this study suggest that the similarity between perceptions of actual and preferred environment could be as important as individualization *per se* in predicting students' achievement of important affective and cognitive aims. This research has interesting practical implications, but one must be careful to ensure that the implications drawn are consistent with the unit of statistical analysis used. It cannot be assumed that an individual student's achievement would be improved by moving him or her to a classroom that matched his or her preferences. Rather, the practical implication of these findings for teachers is that class achievement of certain outcomes might be enhanced by attempting to change the actual classroom environment in ways which make it more congruent with that preferred by the class.

PRACTICAL ATTEMPTS TO IMPROVE CLASSROOM ENVIRONMENTS

In this section we report on how feedback information based on students' perceptions was employed as a basis for the teacher's reflection upon, discussion of, and systematic attempts to improve classroom environments. The basic logic underlying the approach has been described by Fraser (1986) and applied successfully in various studies at the elementary, secondary, and higher education levels (Fraser and Fisher, 1986).

The attempt to improve classroom environment described below made use of the short twenty-four-item version of the CES discussed previously. The class involved in the study consisted of twenty-two Grade 9 boys and girls of mixed ability studying science at a government school in Australia (Fraser and Fisher, 1986). The procedure incorporated the following five fundamental steps:

1. *Assessment.* The CES was administered to all students in the class. The preferred form was answered first, while the actual form was administered in the same time slot one week later.

2. *Feedback.* The teacher was provided with feedback information derived from students' responses in the form of the profiles shown in figure 6-2 representing the class means of students' actual and preferred environment scores. These profiles permitted ready identification of the changes in classroom environment needed to reduce major differences between the nature of the actual environment as perceived by students and the preferred environment as currently perceived by students. Figure 6-2 shows that students would prefer more of all dimensions except Affiliation.

3. *Reflection and Discussion.* The teacher engaged in private reflection and informal discussion about the profiles in order to provide a basis for a decision about whether an attempt would be made to change the environment in terms of some of the dimensions of the CES. The teacher decided to introduce an intervention aimed at increasing the levels of Teacher Support and Order and Organization in the class.

4. *Intervention.* The teacher introduced an intervention of approximately two months' duration in an attempt to change the classroom environment. This intervention consisted of a variety of strategies, some of which originated during discussions among teachers. Others were suggested by examining ideas contained in individual CES items. For example, strategies used to enhance Teacher Support involved the teacher moving around the class more to mix with students, providing assistance to students, and talking with them more than previously. Strategies used to increase Order and Organization involved taking considerable care with distribution and collection of materials during activities and ensuring that students worked more quietly.

5. *Reassessment.* The actual form of the scales was readministered to students at the end of the intervention to see whether they were perceiving their classroom environments differently than before.

The results are summarized graphically in figure 6-2, in which the dotted line indicates the class mean score for students' perceptions of actual environment on each of the five CES scales at the time of posttesting. The graph clearly shows that some change in student perceptions of actual environment occurred during the time of the intervention. Pretest-posttest differences were statistically significant only for Teacher Support, Task Orientation, and Order and Organization. These findings are noteworthy because two of the dimensions on which appreciable changes were recorded were those on which the teacher had attempted to promote change. (Note also that the classroom appears to have become more task oriented than the students would have preferred, possibly a side effect of the intervention.) Overall, this study, in conjunction with previous studies (Fraser and Fisher, 1986), suggests that teachers

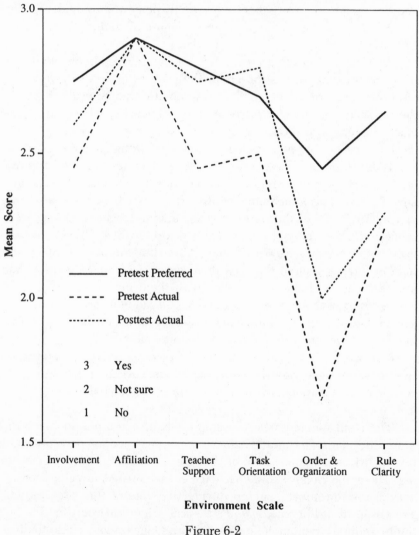

Environment Scale

Figure 6-2

Mean Pretest Actual, Pretest Preferred, and Posttest Actual Profiles

can employ instruments that assess classroom environment to secure information that is useful in improving the environment in their classrooms.

CURRENT TRENDS AND DESIRABLE FUTURE DIRECTIONS

We turn now to some of the new lines of research on educational environments that have implications for the improvement of science education.

Combining Qualitative and Quantitative Methods

Although only limited progress has been made toward the desirable goal of combining quantitative and qualitative methods within the same study in research on classroom learning environments, the value of a confluence of qualitative and quantitative methods is illustrated in detail in the studies reported below. For example, a team of thirteen Australian researchers was involved in over 500 hours of intensive classroom observation of twenty-two exemplary teachers and a comparison group of nonexemplary teachers (Fraser and Tobin, 1989). The main methods for data collection involved classroom observation, interviews with students and teachers, and the construction of case studies. But a distinctive feature was that the qualitative information thus obtained was complemented by quantitative information obtained from questionnaires assessing students' perceptions of classroom psychosocial environment. These instruments furnished a useful picture of life in exemplary teachers' classrooms as seen through the students' eyes. The results from use of the qualitative and quantitative data provided considerable evidence (a) that exemplary and nonexemplary teachers can be differentiated in terms of the psychosocial environments of their classrooms as perceived by their students and (b) that exemplary teachers typically create and maintain environments that are markedly more favorable than those of nonexemplary teachers (Fraser and Tobin, 1989). For example, relative to a comparison group of students of nonexemplary teachers, the students of an exemplary biology teacher perceived their class as having a much more favorable environment, especially in terms of Involvement, Teacher Support, and Order and Organization.

In another study, which focused on the elusive goal of higher-level cognitive learning, a team of six researchers studied intensively the

Grade 10 science classes of two teachers (Peter and Sandra) over a ten-week period (Tobin, Kahle, and Fraser, 1990). Each lesson was observed by several researchers, students and teachers were interviewed daily, and students' written work was examined. The study also involved quantitative information from questionnaires assessing students' perceptions of classroom psychosocial environment. An important finding was that students' perceptions of the learning environment within each class were consistent with the observers' field records of the patterns of learning activities and engagement in each classroom. For example, the high level of Personalization perceived in Sandra's classroom matched the large proportion of time that she spent in small-group activities during which she constantly moved about the classroom interacting with students. The lower level of Personalization perceived in Peter's class was associated partly with the larger amount of time spent in the whole-class mode and the generally public nature of his interactions with students.

School Psychology

Given the school psychologist's increasing responsibility for affective as well as cognitive aspects of schooling, the field of classroom psychosocial environment provides a good example of an area that furnishes a number of ideas, techniques, and research findings which could be valuable in school psychology. Traditionally, school psychologists have tended to concentrate heavily and sometimes exclusively on their roles in assessing and enhancing academic achievement and other valued learning outcomes. The study of classroom environment provides an opportunity for school psychologists and teachers to become sensitized to subtle but important aspects of classroom life. For example, Burden and Fraser (in press) report the way in which school psychologists used instruments to assess classroom environment in helping British teachers to change their classroom interactive styles, and in using discrepancies between students' perceptions of actual and preferred environment as an effective basis to guide improvements in their classrooms.

Constructivist Learning Environments

Traditionally, teachers have conceived their roles to be concerned with revealing or transmitting the logical structures of their knowledge, and directing students through rational inquiry toward discovering predetermined truths expressed in the form of laws, principles, rules, and algorithms. Recent developments in history, philosophy, and sociology have provided educators with a better understanding of how knowledge is acquired. For the individual learner, meaningful learning is seen as a cognitive process of making sense of the individual's new experiences in relation to the totality of the individual's already constructed knowledge. Because the individual belongs to a world populated by significant others, the sense-making process involves active negotiation and consensus building (see chapter 3 in this volume).

A new learning environment instrument was developed to help researchers and teachers assess the degree to which a particular classroom environment is consistent with a constructivist epistemology, and to help teachers reflect on their epistemological assumptions and reshape their teaching practice. This instrument is called the *Constructivist Learning Environment Survey* (*CLES*) (Taylor, Fraser, and White, 1994). Because it is desirable that the CLES empowers teachers to overcome constraints on the development of constructivist learning environments, a critical theory perspective was incorporated into the instrument. The scales in the CLES are Personal Relevance (the relevance of classroom experience to out-of-school experiences); Student Negotiation (emphasis on creating opportunities for students to explain and justify their ideas, and to test the viability of their own and others' ideas); Uncertainty (provision of opportunities for students to experience knowledge as arising from inquiry, as involving human experience and values, as evolving and insecure, and as culturally and socially determined); Shared Control (opportunity for students to negotiate and share control of learning activities, assessment, and social norms); and Critical Voice (emphasis on students' questioning the prevailing curriculum, teaching methods, and assessment policy).

Researchers could use the CLES in (a) monitoring the effectiveness of preservice/in-service attempts to change teaching/learning styles to a more constructivist approach; (b) evaluating the impact of constructivist approaches to teaching on student outcomes; (c) guiding attempts by the teacher as researcher to reflect upon and improve classroom environment; (d) reducing the amount of classroom observation needed in studies of constructivist teaching and learning (by collecting information from students via the CLES); (e) complementing qualitative information by constructing richer case studies that also include quantitative information based on students' perceptions obtained with the CLES; and (6) investigating the relationship between teacher cognition and teaching practice.

Personal Forms of Scales

There is a major problem with nearly all existing classroom environment instruments when they are used to identify differences between subgroups within a classroom (for example, boys and girls) or in the construction of case studies of individual students. The problem is that items are worded in such a way that they elicit an individual student's perceptions of the class as a whole, as distinct from that student's perceptions of his or her own role within the classroom. For example, items in the traditional "class" form of classroom environment instruments might seek students' opinions about whether "the work of the class is difficult" or whether "the teacher is friendly toward the class." In contrast, a "personal" form of the same items would seek opinions about whether "*I* find the work of the class difficult" or whether "the teacher is friendly toward *me*."

A vivid example of the way in which certain subgroups of students within a science class perceived different subenvironments because of the teacher's differential treatment of them is provided by a study of targeted students, in this case, pupils who monopolized the verbal interaction during whole-class activities (Fraser, 1994). It was found that targeted students perceived significantly greater levels of involvement and rule clarity than nontargeted students, which was consistent with classroom observations showing that the teachers directed more questions at targeted students and allowed them (and not other students) to call out answers without being asked. Similarly, in another study combining qualitative and quantitative methods (Tobin, Kahle, and Fraser, 1990), case studies of individual students revealed that meaningful differences in perceptions of classroom environment existed between certain students, and that those differences were consistent with the teacher's expectations of and attitudes toward individuals. The findings of these two studies highlight the need for a new generation of instruments for assessing classroom environment that are more capable of detecting the differences in perceptions between individuals or subgroups within the class.

Fraser, McRobbie, and Giddings (in press) have developed and validated parallel "Class" and "Personal" forms of both an actual and preferred version of the *Science Laboratory Environment Inventory* (*SLEI*). The version of the SLEI shown in figure 6-1 is the "personal actual" form.

Three uses of the new personal form of the SLEI were reported by Fraser, McRobbie, and Giddings (in press). First, students' scores on the class form were found to be systematically more favorable than their scores on the personal form, perhaps suggesting that students have a more detached view of the environment as it applies to the class as a whole. Second, an investigation of gender differences in student perceptions of science/laboratory classes suggested that, as hypothesized, gender differences in perceptions were somewhat larger on the personal form than on the class form. Third, although a study of associations between student outcomes and their perceptions of the science laboratory environment revealed that the magnitudes of associations were comparable for class and personal forms of the SLEI, commonality analyses showed that each form accounted for appreciable amounts of outcome variance which was independent of that explained by the other form. This finding serves to justify the decision to have separate class and personal forms because they do appear to measure different, albeit overlapping, aspects of the classroom environment in the science laboratory.

Incorporating Learning Environment Ideas into the Education of Science Teachers

The improvement of preservice and in-service education programs for science teachers requires the input of new ideas which will help teachers become more reflective and retrospective about their teaching. Despite the fact that the thriving field of psychosocial learning environment furnishes a number of ideas and techniques which are potentially valuable for inclusion in teacher education programs, surprisingly little progress has been made in incorporating such ideas into those programs. Fisher and Fraser (1991) reviewed some examples of successful past and current attempts to include studies of learning environment in teacher education, and made suggestions about how teacher education in the future can be improved through the input of ideas from research on learning environment.

In particular, Fisher and Fraser (1991) reported some case studies of how classroom and school environment work has been used in preservice and in-service teacher education to (a) sensitize teachers to subtle but important aspects of classroom life, (b) illustrate the usefulness of including classroom and school environment assessments as part of a teacher's overall evaluation/monitoring activities, (c) show how assessment of classroom and school environment can be used to facilitate practical improvements in classrooms and schools, and (d) provide a valuable source of feedback about teaching performance in formative and summative evaluations of student teaching. It appears that information on student perceptions of the classroom learning environment during preservice teachers' field experience adds usefully to the information obtained from university supervisors, school-based cooperating teachers, and student teacher self-evaluation.

CONCLUSIONS AND IMPLICATIONS

In less than thirty years, older classroom environment instruments have been used more widely and cross-validated in various countries, "preferred" forms have been developed to augment the original "actual" or "experienced" forms, short and hand-scorable forms have been designed for the convenience of teachers, and new instruments have been developed to fill gaps (for example, for use in higher education classrooms or science laboratory classes). Workers around the world are continuing to translate and adapt instruments for use in different countries, to develop new instruments for settings in which existing questionnaires are not suitable (for example, computer-assisted instruction, preschool classrooms, constructivist classrooms), and to use the instruments in settings where they have not been previously used, such as special education classes. Some of the many current and promising recent lines of research on educational environments include the use of a "Personal" (as well as a "Class") form of the questionnaires, and the establishment of links between environments (especially the school and the classroom). Also, the topic of classroom environment is beginning to be included in preservice and in-service courses for teachers around the world, and is gaining attention among school psychologists.

Our major purpose in this chapter has been to make this research tradition in science education more accessible to wider audiences. On the basis of work discussed in this chapter we see the following implications for improving science education.

1. Because measures of learning outcomes alone cannot provide a complete picture of the educational process, assessments of learning environment should also be used to provide information about subtle but important aspects of classroom life.

2. Because teachers and students have systematically different perceptions of the learning environments of the same classrooms (the "rose-colored glasses" phenomenon), feedback from students about classrooms should be collected when evaluating the performance of practice teachers during their field experience and when investigating staff development programs.

3. Science teachers should strive to create "productive" learning environments as identified by research. Cognitive and affective outcomes are likely to be enhanced in classroom environments characterized by greater organization, cohesiveness, and goal direction and by less friction. In laboratory classroom environments specifically, greater integration between practical work and the theoretical components of a course tends to lead to improved student outcomes.

4. In order to improve student outcomes, classroom environment should be changed to make it more similar to that preferred by the class.

5. The evaluation of innovations and new curricula should include classroom environment instruments to provide economical, valid, and reliable process measures of effectiveness. For example, the Science Laboratory Environment Inventory listed in figure 6-1 could be used in evaluating innovative approaches to laboratory teaching, whereas the Constructivist Learning Environment Survey could be used to evaluate the many recent attempts to change science classrooms to a more constructivist orientation (see chapter 3 in this volume).

6. Teachers should use assessments of their students' perceptions of actual and preferred classroom environment to monitor and guide attempts to improve classrooms. The broad range of instruments available enables science teachers to select a questionnaire or particular scales to fit their personal circumstances.

7. When assessing classroom environment, a combination of qualitative and quantitative methods should be used instead of either method alone.

8. When one is interested in differences between subgroups of students within a classroom (for example, boys and girls) or in individual students, use should be made of "personal" rather than "class" forms of learning environment instruments.

9. Learning environment assessments should be used by school psychologists in helping science teachers change their interactive styles and improve their classroom environments.

Student outcomes have typically been studied using quantitative approaches based on traditional educational measurements, whereas classroom processes or environment usually have involved qualitative approaches such as informal observations and interviews. We have shown that classroom climate also is susceptible to quantitative study. Admittedly quantitative measures have well-known limitations, but so too do qualitative approaches. Rather than claiming that quantitative methods are superior to qualitative ones in the study of classroom environments, our intention has been to make a potentially useful tradition of quantitative assessment of classroom climate readily accessible so that studies might benefit from the use of a range of quantitative and qualitative approaches.

We assume in this chapter that having a positive classroom environment is an educationally desirable end in its own right. Moreover, the comprehensive evidence presented here also clearly establishes that the nature of the classroom environment has a potent influence on how well students achieve a range of desired educational outcomes. Consequently, science educators need not feel that they must choose between striving to achieve positive classroom environments and attempting to enhance students' achievement of cognitive and affective aims. Rather, positive educational climates can be viewed as both means to valuable ends and as worthy ends in their own right.

REFERENCES

Burden, Robert, and Fraser, Barry J. "Use of Classroom Environment Assessments in School Psychology: A British Perspective," *Psychology in the Schools*, in press.

Docker, John G.; Fraser, Barry J.; and Fisher, Darrell L. "Differences in the Psychosocial Work Environment of Different Types of Schools," *Journal of Research in Childhood Education* 4 (1989): 5-17.

Fisher, Darrell L., and Fraser, Barry J. "A Comparison of Actual and Preferred Classroom Environment as Perceived by Science Teachers and Students," *Journal of Research in Science Teaching* 20 (1983): 55-61.

Fisher, Darrell L., and Fraser, Barry J. "Incorporating Classroom Environment Ideas into Teacher Education: An Australian Perspective." Paper presented at the Annual Meeting of the American Educational Research Association, Chicago, IL, 1991.

Fraser, Barry J. *Classroom Environment*. London, England: Croom Helm, 1986.

Fraser, Barry J. *Individualized Classroom Environment Questionnaire*. Melbourne, Victoria: Australian Council for Educational Research, 1990.

Fraser, Barry J. "The Learning Environment in Science Classrooms and its Effects on Learning." Paper presented at the International Conference on Science Education in Developing Countries: From Theory into Practice, Jerusalem, Israel, 1993.

Fraser, Barry J. "Research on Classroom and School Climate." In *Handbook of Research on Science Teaching and Learning*, edited by Dorothy Gabel, pp. 493-541. New York: Macmillan, 1994.

Fraser, Barry J.; Anderson, Gary J.; and Walberg, Herbert J. *Assessment of Learning Environments: Manual for Learning Environment Inventory (LEI) and My Class Inventory (MCI)*, 3d version. Perth, Western Australia: Western Australian Institute of Technology, 1982.

Fraser, Barry J., and Fisher, Darrell L. "Predicting Students' Outcomes from Their Perceptions of Classroom Psychosocial Environment," *American Educational Research Journal* 19 (1982): 498-518.

Fraser, Barry J., and Fisher, Darrell L. "Student Achievement as a Function of Person-Environment Fit: A Regression Surface Analysis," *British Journal of Educational Psychology* 53 (1983): 89-99.

Fraser, Barry J., and Fisher, Darrell L. "Using Short Forms of Classroom Climate Instruments to Assess and Improve Classroom Psychosocial Environment," *Journal of Research in Science Teaching* 23 (1986): 387-413.

Fraser, Barry J.; McRobbie, Campbell J.; and Giddings, Geoffrey J. "Evolution and Validation of a Personal Form of an Instrument for Assessing Science Laboratory Classroom Environment," *Journal of Research in Science Teaching*, in press.

Fraser, Barry J., and O'Brien, Peter. "Student and Teacher Perceptions of the Environment of Elementary-School Classrooms," *Elementary School Journal* 85 (1985): 567-580.

Fraser, Barry J., and Tobin, Kenneth. "Student Perceptions of Psychosocial Environments in Classrooms of Exemplary Science Teachers," *International Journal of Science Education* 11 (1989): 19-34.

Fraser, Barry J., and Treagust, David F. "Validity and Use of an Instrument for Assessing Classroom Psychosocial Environment in Higher Education," *Higher Education* 15 (1986): 37-57.

Fraser, Barry J., and Walberg, Herbert J., eds. *Educational Environments: Evaluation, Antecedents, and Consequences*. Oxford, England: Pergamon Press, 1991.

Fraser, Barry J.; Williamson, John C.; and Tobin, Kenneth. "Use of Classroom and School Climate Scales in Evaluating Alternative Schools," *Teaching and Teacher Education* 3 (1987): 219-231.

Goldstein, Harvey. *Multilevel Models in Educational and Social Research*. London, England: Charles Griffin, 1987.

Haertel, Geneva D.; Walberg, Herbert J.; and Haertel, Edward H. "Socio-Psychological Environments and Learning: A Quantitative Synthesis," *British Educational Research Journal* 7 (1981): 27-36.

Halpin, Andrew W., and Croft, Don B. *Organizational Climate of Schools*. Chicago, IL: Midwest Administration Center, University of Chicago, 1963.

Moos, Rudolf H. *The Social Climate Scales: An Overview*. Palo Alto, CA: Consulting Psychologists Press, 1974.

Moos, Rudolf H., and Trickett, Edison J. *Classroom Environment Scale Manual*, 2d ed. Palo Alto, CA: Consulting Psychologists Press, 1987.

Stern, George G. *People in Context: Measuring Person-Environment Congruence in Education and Industry*. New York: Wiley, 1970.

Taylor, Peter C.; Fraser, Barry J.; and White, Loren. "CLES: An Instrument for Monitoring the Development of Constructivist Learning Environments." Paper presented at the Annual Meeting of the American Educational Research Association, New Orleans, LA, 1994.

Tobin, Kenneth; Kahle, Jane B.; and Fraser, Barry J., eds. *Windows into Science Classrooms: Problems Associated with Higher-Level Cognitive Learning*. London, England: Falmer Press, 1990.

Welch, Wayne W., and Walberg, Herbert J. "A National Experiment in Curriculum Evaluation," *American Educational Research Journal* 9 (1972): 373-383.

Wubbels, Theo; Brekelmans, Mieke; and Hooymayers, Herman P. "Interpersonal Teacher Behavior in the Classroom." In *Educational Environments: Evaluation, Antecedents, and Consequences*, edited by Barry J. Fraser and Herbert J. Walberg, pp. 141-160. Oxford, England: Pergamon Press, 1991.

Wubbels, Theo; Brekelmans, J. Mieke G.; and Hooymayers, Herman P. "Do Teacher Ideals Distort the Self-Reports of Their Interpersonal Behavior?" *Teaching and Teacher Education* 8, no. 1 (1992): 47-58.

Wubbels, Theo, and Levy, Jack. "A Comparison of Interpersonal Behavior of Dutch and American Teachers," *International Journal of Intercultural Relations* 15 (1991): 1-18.

Chapter 7

TEACHER CHANGE AND THE ASSESSMENT OF TEACHER PERFORMANCE

Kenneth Tobin

In the past twenty years, few challenges have been as great as those associated with educating and maintaining a viable work force of teachers. Included among the most formidable challenges have been problems associated with teacher education and assessing teachers' performance. Interestingly, questions associated with each of these problem areas relate to the nature of teacher knowledge, the manner in which teacher knowledge is constructed and reconstructed, and how knowledge is represented in action. Not surprisingly, there have been diverse ways of framing and answering such questions and these have profound implications for the practice of education.

For the greater part of this twenty-year period, the dominant epistemology underlying research and development in the domains of teacher education and teacher assessment has been *objectivism*. An assumption of objectivism is that knowledge can be regarded as an entity which exists independently of knowers, to be experienced in an objective way, and learned by teachers who then can represent it validly in the sense that the knowledge-in-action matches the entity that was learned. Similarly, the products of research were regarded as truths to be learned and adopted by teachers. These truths represented the knowledge base of teaching and were incorporated into curricula for prospective and practicing teachers and into systems of teacher assessment. The products

Chapter consultants: Tom Russell (Queen's University, Canada), Angelo Collins (Florida State University, USA), and James Gallagher (Michigan State University, USA).

from research thereby became criteria for teacher training and bases for ascertaining whether teaching performance was effective.

To an increasing extent, educators have raised questions about the applicability of objectivism to the framing and resolution of problems associated with teacher education and teacher assessment. *Constructivism*, which is an alternative set of beliefs about knowing, offers the potential of addressing problems in a different manner. According to the constructivist perspective, a reality exists independently of knowers, but that reality can only be experienced subjectively. That is, knowers construct models of reality to fit their experiences and extant knowledge. The search for absolute truths will be elusive forever. The test of the personal truth of knowledge rests in its viability, that is, the extent to which an individual is able to pursue his or her goals in a social milieu. In the context of teacher education, it is important to realize that, rather than one best method of teaching and learning, there will be only strategies that prove to be viable for individuals in the circumstances in which the curriculum is implemented. The challenge for teacher educators, then, is to build programs that enable teachers to learn what they need to know in order to implement the curriculum in such a way that student learning is improved. This implies changes from past and current practices.

Lambert (1988) noted that in-service and preservice activities and programs have served to reinforce the status quo rather than change it, thus perpetuating a paternalistic system that reinforces traditional practices. Lieberman and Miller (1990) stated that in-service education has come to be synonymous with training and implies a deficit model of education. In contrast, they proposed the concept of teacher development to describe a new approach to professional growth activities that allows teachers to build personal learning on a foundation of tacit knowledge through ongoing inquiry and analyses. There is more to the differences in approach than the semantics associated with the terms "in-service training" and "professional growth activities." Embodied in the latter approach is a belief that, as a result of conducting research in their own classrooms, teachers can construct knowledge that is worthwhile. Furthermore, Lieberman and Miller proposed that knowledge constructed in this manner is more likely to lead to solutions to today's educational problems.

Habermas's theory of "knowledge constitutive interests" connects knowledge with the interests of individuals or sets of individuals. Grundy

(1987) applied the theory to curriculum in order to draw attention to the significance of power relationships between teachers and learners and the epistemologies embedded within activities. A curriculum that emphasizes *technical interests* is characterized by rote learning of facts and adherence to rules and procedures. Control and power usually reside with a teacher or someone other than the learner. The epistemology underpinning technical interests is objectivism. In contrast, constructivism is the epistemology underlying practical and emancipatory interests. A curriculum characterized by *practical interests* provides learners with opportunities to interact and negotiate understanding of content. *Emancipatory interests* are distinctive in that learners have autonomy with respect to their learning.

Grundy's framework is similar to Fien's (1991) categorization of teacher education curricula as "traditionalist," "rationalist," and "critical." According to Fien, traditionalist approaches to teacher education are based on objectivism, relying on the transfer of knowledge from those who have it, such as master teachers and teacher educators, to those who need to learn. Rationalist approaches conceptualize teacher education as an applied science built upon the "best" principles and practices derived from research. In contrast, critical approaches to teacher education encourage reflection on teaching, the social context in which learning and teaching occur, and the interests served by alternative educational practices. Whereas traditional and rationalist approaches emphasize technical interests, the critical approach emphasizes emancipatory interests. That is, teachers are provided with autonomy with respect to what and how they learn. The critical approach is empowering in the sense that teachers are regarded as professionals who must be able to function in an environment that requires the resolution of complex ethical dilemmas. Consequently, teachers have the autonomy to identify problems in their classrooms and seek solutions that make sense to them. The processes of teacher education and curriculum reform can be seen as complementary.

Hart and Robottom (1990) noted that curriculum development and professional development need to be regarded as interactive elements of the same reform process with each being seen as problematic and amenable to change. They advocated a method of curriculum reform that:

. . . is critical of the conditions creating theory/practice discontinuities and . . . empowers teachers as researchers and participants in the decision-making process. That is, the reform process should be informed by critical reflection on the relationship between theory and practice and educational situation. [P. 584]

New approaches to conceptualizing teacher education, such as those described above, have the promise of reforming school curricula dramatically. However, in association with a new conceptualization of teacher education, there are emergent questions which demand convincing answers. The economic downturn during the past few years has increased the necessity of accountability for the huge financial investments in education. One significant focus of accountability is the extent to which teachers demonstrate competence in the classroom. However, within the objectivist paradigm, assessing teacher performance has not been done well and, from a constructivist perspective, the challenges seem even greater because existing assessment systems seem grossly inadequate.

In this chapter, I examine teacher change in terms of teacher learning and factors associated with implementing a curriculum. Whereas teacher education has a critical role in introducing and sustaining changes, recognition is given to the importance of the sociocultural contexts in which changes are to occur. The emergence of professional practice schools (that is, schools that are involved in the education of teachers and prospective teachers) has been one response to the need to take account of contextual factors that often inhibit sustained reform. However, it is essential that changes in the practices of teachers and students are monitored. Credible measures of teaching performance can be used to ascertain whether or not teacher education programs make a difference in the classroom and to identify teaching practices that might need to be changed. In the following sections I examine approaches to teacher education, sociocultural perspectives on change, the use of professional practice schools to facilitate learning and change, and the assessment of teaching performance. Emphasis is given to the importance of the commensurability of the practices adopted in each of these interrelated areas.

TEACHER EDUCATION

Promising innovations in teacher education, such as strategy analysis and coaching, have fallen short in terms of preparing teachers to introduce and sustain educational reform. The approach advocated here, however, views teachers as professionals who, given the autonomy to maintain a central role in their own educational programs, engage in reflective actions oriented toward identifying and retaining viable classroom practices and changing strategies that do not work as intended. The approaches underlying strategy analysis and coaching are quite different from the reflective practices advocated to promote teacher learning as a basis for change. The research reviewed here reflects changes in thinking about teacher education and associated research over a period of approximately twenty years.

Strategy Analysis

In the latter part of the 1970s and early 1980s, researchers sought to identify teaching practices associated with higher achievement. It was assumed that, once identified, these optimal practices could be used by others to improve student achievement. Accordingly, models consisting of sets of teaching behaviors were developed and became the basis for teacher training programs. Models were learned, practiced in micro and peer groups, and then applied in classrooms. For example, Yeany (1977) reported a change in teaching style and attitudes when preservice teachers used a model to analyze micro-lessons systematically. He noted that provision for interaction with an instructor, as well as analysis of teaching, produced the largest changes.

In a meta-analysis of research on changing teaching behavior, Yeany and Porter (1982) described five types of treatment that had been used: study of an analysis system, self-analysis of recorded lessons, observation of model lessons, analysis of model lessons, and feedback guided by an analysis system. Feedback guided by analysis was the most effective method of changing teacher behavior, whereas analysis of written models was the least effective. The results suggested that all methods were effective in improving teaching performance. Additional results from the Yeany and Porter meta-analysis indicated that microteaching produced better results than no microteaching, analyzing results was better than only observing model lessons, using an analysis system was better than not using one, and models involving analysis of videotapes and audiotapes were better than ones involving written models. Each of these findings is consistent with the interpretation that greater changes in teacher behavior were associated with intervention strategies that provided more opportunities for active cognitive processing.

Strategy analysis relies on an expert to decide what teachers ought to know, represent that knowledge in a model, and then train teachers to acquire and implement faithfully the elements of the model. Although teachers ought to understand the theories underlying the model, they are expected to implement the model with fidelity. Such approaches are open to Lieberman and Miller's (1990) charge of assuming a deficit model of teacher learning and change. Furthermore, there is no formal attempt to build on the tacit knowledge of teachers or to raise questions about the applicability of the model to local circumstances.

Coaching

An extension of strategy analysis research involved coaching. Joyce and Showers (1983) noted that teachers could learn to improve almost any aspect of teaching as long as adequate training was provided. The chief problem was integrating a new skill into existing patterns of behavior. Four elements comprise an in-service model with a planning and implementation focus: the study of the theoretical basis or the rationale of the teaching method; the observation of demonstrations by persons who are relatively expert in the teaching method; practice and feedback in relatively protected conditions; and coaching one another to facilitate incorporation of the new method into the day-to-day teaching style. Joyce and Showers defined coaching in terms of the provision of on-site personal support and technical assistance for teachers. The role of the coach is to assist teachers in analyzing teaching situations and to enable a strategy to be used appropriately in the class according to the curriculum objectives and student abilities. The coach also provides a psychological support system to encourage teachers to practice in the face of occasional failures.

Joyce and Showers differentiated between mastering a skill in a workshop setting and transferring the skill to the everyday classroom. They demonstrated that continuous practice, feedback, and the companionship of coaches are essential to enable even highly motivated teachers to bring additions to their teaching repertoire under control. Joyce and Showers stated that, in order for a teacher to adopt a new strategy, the teacher first should develop skill in the strategy. This level of understanding can be achieved through involvement in the kinds of workshops and courses that traditionally have been offered. Next, the teacher needs to attain executive control over the strategy and adapt the skill to the classroom setting and students. Horizontal transfer is said to have occurred when the skill is transferred directly from the training situation to the workplace. Vertical transfer implies a higher level of learning in which the teacher has adapted the skill to be applicable in specific contexts and for specific students.

The notions of horizontal and vertical transfer were important contributions of the research on coaching. Vertical transfer introduced the question of appropriateness. In addition, peer coaching recognized the problems that arise when outsiders diagnose problems and provide solutions to be adopted by insiders. However, it soon became apparent that anyone who identified a problem and an apparent solution was likely to be regarded as an outsider.

A series of studies involved investigation of the effectiveness of three peer coaching models with experienced middle and high school science and mathematics teachers. In the first model, each of five high school mathematics teachers coached one another to improve a self-selected teaching strategy (Tobin and Espinet, 1990). At a given time, no teacher coached the person coaching him or her. Each teacher observed a lesson of a colleague and, one week later, there was a one-hour coaching session during which teachers collaborated in analyzing the observed lesson. The second model involved two teachers coaching one another on the same day as the observation. The third model incorporated a coach who was external to the school system, and coaching sessions took place on the same day as the observation. All coaching arrangements were successful in facilitating change for the teachers involved in the study. However, there were some aspects of peer coaching that were less successful than others. Teachers perceived as weak teachers by their peers were perceived to be weak coaches as well. Another problem was that some teachers believed that they had few weaknesses and did not appear to consider seriously suggestions for improvement.

The main advantage of coaching sessions was that they provided a context for reflection on practice. All teachers claimed to benefit from observing other teachers in action. Better teachers appeared able to reflect on their actions by themselves, whereas weaker teachers seemed to need a colleague to provide answers and foci for discussion. However, rather than use the results of research to guide colleagues during coaching sessions, teachers tended to use practical knowledge based on personal classroom teaching experience. Most teachers had little to share after one coaching session.

Learning as a Basis for Teacher Change

Some innovations, such as coaching, promised a new approach to teacher education, but did not yield the favorable results that were envisioned. Little (1989) claimed that most coaching projects only involve changes regarded as superficial, safe, and inconsequential, and hence they have little impact on the school culture. When changes occurred, the most significant effects of coaching seemed to be attributable to collaboration between teachers and coaches. The extent and nature of change were dependent on the persons involved in the collaboration, opportunities to experience alternative images of teaching and learning, and the extent to which discussions led to reflection on classroom practices (Tobin and Espinet, 1990).

When teacher learning is viewed from a constructivist perspective. (von Glasersfeld, 1989), learning is seen as a social process of making sense of experience in terms of extant knowledge. Learning about teaching is accomplished best through the direct experience of the teacher-learner, in conjunction with opportunities to reflect critically on the experience and emergent problems. If the persistent problems associated with teaching and learning are to be solved, it is necessary to pose questions about what is and is not done in classrooms, to seek answers, and, on the basis of those answers, to consider change. Learning is both an interactive and a constructive process. As a consequence, learning occurs in social settings as persons interact to negotiate meanings and arrive at consensus.

A teacher's experience is sensory and is given meaning by reflection that involves the construction of images and, in some cases, the assignment of language to images, which can be thought of as dynamic reconstructions of experience. Sanders and McCutcheon (1986) reported how teachers used images as they thought about teaching. Tobin and Ulerick (1989) observed how a teacher perceived her roles in terms of metaphors and associated visual images. An even more graphic example involved a science teacher who envisioned himself as a swashbuckling captain of a ship, barking orders to his crew to keep them under tight control (Tobin, Kahle, and Fraser, 1990).

Metaphors are at the heart of conceptualization and can be used to make sense of concepts associated with teaching and learning (for example, Munby and Russell, 1990). Although numerous studies have examined the role of metaphor in making sense of teaching, not all authors have conceptualized metaphor from a constructivist point of view. When people learn something new from a constructivist perspective, they make initial sense of it in terms of something that is known already. The initial link is metaphorical.

The possibility that a teaching role might be defined by different metaphors on different occasions was evident when a teacher used the metaphor of entertainer to make sense of his management role on some occasions, while on others perceiving his role as the captain of the ship (Tobin, Kahle, and Fraser, 1990). The classroom environment appeared to be very dependent on the metaphor which the teacher used to define his role in different contexts. When the teacher decided that it was appropriate to be captain of the ship, he maintained strict discipline and was a hard task master. The captain of the ship delivered content, explained areas that were potentially difficult to understand, and ensured that the content was covered in a timely fashion. Knowing the right answers was a high priority in this instructional mode. However, the highest priority was keeping the students on task and ensuring that they obeyed the teacher's orders. There were harsh penalties for not following orders. The teacher's beliefs, associated deliberative actions, and routine practices, embedded within an image and metaphor of being captain of the ship, constrained the roles of students. The students engaged in a range of teacher-centered, whole-class activities which were prescribed by the teacher such as listening to the teacher, answering questions when called upon, providing written responses to assignments.

As an entertainer, the teacher focused on humor. He was intent on projecting an image of being friendly, amenable to interactions with students, and having interests that were not unlike those of the students in the class. The entertainer role was evident in whole-class, small-group, and seatwork activities. In this mode, the teacher told stories about the world outside of the classroom, his personal life outside of education, and other personal, humorous anecdotes. The major difference between the entertainer and captain of the ship metaphors was that, in the former instance, the teacher demonstrated that he was amenable to off-task behavior, and that students could be humorous at his expense. It was safe to take risks with the entertainer. It was also in this mode that gender-related differences were most apparent in the interactions between students and the teacher. With some male students, the teacher demonstrated macho tendencies and with the more attractive females he engaged in social discourse. The entertainer was not in a rush to cover content or to keep students on task. It was a time to enjoy school, to relax, and to wander off task. Engagement was not controlled by the entertainer. As long as students did not become too disruptive, they were free to work at their own pace while they completed exercises from a workbook or the textbook.

There were similarities in the two metaphors as well. Each demonstrated an emphasis on technical interests (Grundy, 1987). The captain of the ship controlled what was to be learned and the entertainer interacted with students while their tasks were controlled by the workbook or textbook. Both metaphors were embedded within objectivism and a myth of the school-as-workplace. Thus, the teacher switched between roles when he perceived the context to be appropriate. The teacher defined his role in the way that made sense to him in the circumstances. Each guiding metaphor or image appears to have been a routine for the teacher. Referring to the metaphor was not a conscious or deliberative process as he taught. However, when he reflected on his teaching, the teacher used the terms captain of the ship or entertainer to describe what he was doing and to justify his thoughts and practices. Furthermore, there was ample evidence of both metaphors in the actions of the teacher as he taught and reflected on teaching.

Russell (in press) encouraged educators to move beyond the everyday meaning given to reflection. He noted that "reflection is related to analysis of the nature of professional knowledge and the ways it is acquired, held, and renewed, and it is driven by perspectives on the relationship between thought and action" (p. 1). Russell emphasized the need to recognize that learning can be constructed from experience, and that reflection cannot be separated from the "immediacy of practice." According to Russell, "reflection-in-action involves reframing of experience in response to messages from the practice setting, followed by new actions in the practice setting, actions that express the new frame and test its viability" (p. 8). Reflection involves deliberating on reconstructed knowledge. What emerges as having important implications for teacher education is the range of objects for reflection. It is possible that teachers can identify and effect salient changes by reflecting on images, meta-phors, beliefs, and values in relation to what they perceive to be happen-ing in their classrooms. Because metaphors sometimes are used to make sense of what teachers do in classes, it is important for teachers to reflect on their images of teaching and the language which they use to describe teaching and learning. If metaphors constrain the actions of teachers, it is desirable that teachers are conscious of their metaphors and consider alternatives and the consequences of adopting alternative conceptualiza-tions of teaching roles. Reflecting in this manner could enable teachers to modify their visions of what the curriculum could be like and to com-pare what is happening in their classes to the vision of what they would like to happen.

The research reviewed so far does not address the issue of whether teachers prefer to find out for themselves or to be told what to do. Although teachers have to make sense of what they find out or are told, the issue of expectation is important when it comes to learning. Belenky, Clinchy, Goldberger, and Tarule (1986) drew attention to the manner in which women learn and make sense of their experiences. Women tended to be "received knowers" with little or no independent voice, and were likely to accept as valid ideas suggested by others, particularly per-sons perceived as experts. That is, many women preferred to be told what to do rather than work it out for themselves. In contrast, "con-structed knowers" assumed responsibility for working things out for themselves, gave voice to their ideas, and were unlikely to accept knowl-edge in a passive manner from authorities or colleagues. Teachers can be constructed knowers in one context and received knowers in others.

The curriculum for teachers should have an emphasis on practical and emancipatory interests (Grundy, 1987), thus allowing teachers to learn what they find to be of relevance in their own classrooms, schools, and school districts. In this respect, the curriculum ought to empower teachers to assume responsibility for the education of their students. However, empowerment is a process that requires learning to occur. Ini-tially, many teachers will reject the notion that they can or should have such responsibility. Furthermore, many might expect to be told what it is that they need to know in order to solve the problems of education. Par-ticular problems might be anticipated with respect to meeting the expectations of teachers who are inclined to be received knowers in most situations. If a program is planned on the assumption that teachers are constructed knowers, there might be a high level of frustration among received knowers who simply want to be told what to do. This frustration might lead to cynicism with respect to the lack of expertise being shown by educators responsible for the teacher education curriculum.

CULTURAL PERSPECTIVES ON SCHOOL CHANGE

A number of writers have argued that educational reform, or sustained curriculum change, can be achieved only if substantive changes occur in the culture in which the curriculum is embedded. Sarason (1982) concluded that failure to change what happens in schools is due in part to a failure to recognize the power of culture. Within that culture, teachers relate possible actions to referents, which are beliefs and images, to make sense of what they think and do. The customs and taboos that characterize a culture derive from socially negotiated referents, referred to as *myths*, which are reinforced by sanctions for legitimate and illegitimate actions. Britzman (1991) noted that:

Cultural myths provide a set of images, definitions, justifications, and measures for thought, feelings, and agency that work to render as unitary and certain the reality it seeks to produce. Myths provide a semblance of order, control, and certainty in the face of uncertainty and vulnerability of the teacher's world. [P. 222]

Britzman explained that some myths serve as a framework to repress certain notions of pedagogy, while others facilitated the generating of alternative images of teaching and learning. An important myth that has characterized teaching is objectivism, an idea that knowledge is "out there" in the world and only has to be accessed by the senses to be transferred to the learner. Many traditional classroom practices and curricular resources have been built on the myth of objectivism. Accordingly, teachers, students, and policymakers make sense of what they do in terms of objectivism. Innovations are planned and implemented within that traditional framework without questioning the myth. One problematic consequence of adopting objectivism as a referent for educational activities is that the focus is not on the learner as much as it is on aspects of the milieu in which learning is assumed to be embedded.

The teacher as controller of students is a myth that pervades classrooms. Although it does not seem appropriate for students to be emancipated in

every sense, there is a strong rationale for teachers to emphasize emancipatory interests with respect to learning. From a constructivist perspective, students must have significant and meaningful control of their
own learning. Referring actions to the myth of teacher as controller of
students has led to the highly controlled learning environments that
characterize many classrooms.

The traditional culture of schools seems to be permeated by two
interconnected myths: school-as-workplace and teacher as controller of
students. Adherence to the myth of the school-as-workplace is evident in
the management of schools within the traditional culture. Teachers manage classrooms in ways that maintain control of student thinking and
behavior. Rather than placing responsibility for behavior and learning on
the student, teachers arrange students so that they cannot interact with
one another and they assign tasks that keep students busily engaged in
activities of the teacher's choosing. Keeping busy and completing tasks
are regarded as desirable and often are rewarded in the evaluation procedures that are implemented in the traditional culture. Just as control is
emphasized in the sense that employers and shop stewards control
employees, teachers control students in traditional classrooms. Management is seen as the first consideration, and those newly initiated into the
culture are soon exposed to the conventional wisdom of gaining control
in the first few days. Conventional wisdom among the teaching profession suggests: "First gain control over the students. If you don't manage
to do this in the first few days, the remainder of the year will be a disaster." As a consequence, the initial focus is on management: students are
arranged and supervised with control as a major goal. Only when management concerns have been addressed are the learning needs of students given consideration. The teacher decides at what pace students
will cover content, assessment tasks emphasize engagement in and completion of tasks rather than learning, and extrinsic rewards are used to
increase time on task. This type of approach to the curriculum has been
described by Grundy (1987) as emphasizing technical interests.

Constraints can act as myths for a culture. Examples would include
insufficient time and resources to implement the curriculum in the
desired manner, a need to prepare students to succeed on centrally
administered achievement tests, and restrictive policies of the district
and state. In such cases, failure to utilize a constraint as a referent might
result in classroom practices that are taboo within the culture. Thus, in a

context of curriculum change, it is important for the role of constraints to be recognized. Teachers might believe that change is necessary, know what needs to be changed, have a commitment to personal change, and still believe that change should not occur in the particular circumstances.

Three conditions appear to be necessary in order to understand school change: change must be initiated within schools; change involves individual practitioners reflecting on their beliefs; and change must be examined in the context of organizational factors that comprise structural, institutional, and cultural elements. A model for teacher learning and change in school-based contexts can be developed. Three cognitive, nonlinear, mutually interactive requisites for teacher learning and change are identified: to construct a vision of what science classes could be like and to personalize that vision; to construct a commitment to personal change; and to reflect on thoughts in relation to actions and vice versa.

Although teachers might have learned a great deal about teaching and learning, they might not feel inclined to try to make changes within a school unless they perceive the climate to be conducive to change. For this to be the case, it is imperative that collegial and administrative support are available to assist teachers to facilitate learning in a way consistent with their professional knowledge. In such circumstances, it is possible that teacher learning can be translated into a changed curriculum. Without a threshold level of support from within the school, a teacher normally would be inclined to implement the curriculum in accordance with school customs and taboos.

PROFESSIONAL PRACTICE SCHOOLS

Although there has not been a great deal of research to support professional practice schools, their use is consistent with research reviewed in this chapter. A professional practice school provides an environment in which teachers can learn about teaching and learning. A professional practice school aims to be a community of learners in which teachers are committed to learning as an ongoing activity and assume responsibility, not only for their school students, but also for the learning of prospective and practicing teachers who are a part of the school culture. Lieberman and Miller (1990) noted that in a professional practice school:

. . . teaching and learning are interdependent, not separate functions. In this view teachers are learners. They are problem-posers and problem-solvers; they are researchers, and they are intellectuals engaged in unravelling the learning process for themselves and for young people in their charge. Learning is not consumption; it is knowledge production. Teaching is not performance; it is facilitative leadership. Curriculum is not given; it is constructed empirically, based on the emergent needs and interests of learners. Assessment is not judgment; it documents progress over time. Instruction is not technocratic; it is inventive, craftlike, and above all an imperfect human enterprise. [P. 112]

In addition to the K-12 curricula in a professional practice school, there are curricula for the education of prospective and practicing teachers. This curriculum is grounded in the experiences of teaching, observing teaching, and reflecting-in-action.

The goal of educating teachers in professional practice schools is empowerment. In contrast to in-service models of teacher learning and change, which are based on a deficit model of learning, the curriculum for teachers in professional practice schools focuses on emancipatory interests. However, teachers do not have sole responsibility for the curriculum, which must be negotiated with others in the school and with educators from outside the school (for example, university educators). Unless there is a collaborative effort with outsiders, there is a possibility that the perturbations which initiate learning, curriculum change, and cultural reform will not be ongoing.

Leadership by administrators and teachers is required to assist the school in accomplishing its goals by maintaining administrative structures that ensure that learning is given the highest priority and that teaching roles become more facilitative and inquiring. Of particular importance is the need for school administrators to be supportive of teacher learning, curriculum change, and school change. If a professional practice school is to accomplish its mission, it is necessary for the traditional culture of schools to change and for the school's administrative staff to retain active leadership and facilitative roles.

Numerous studies have shown that there are benefits in having teachers collaborating in interaction groups (Tobin, Davis, Shaw, and Jakubowski, 1991). The curriculum can be enhanced if teachers arrange themselves in groups of six to eight people to discuss research findings and factors which influence learning and teaching. The number of collaborating teachers within the group should be large enough to provide diverse perspectives on learning and teaching, thus providing an optimal learning environment characterized by negotiation and consensus building. Group size, however, should not be too large because it is imperative that all teachers have a voice in interaction groups. The frequency of group meetings and other organizational details, such as the agenda for group meetings, are matters for the group to decide. However, the general purposes of meetings are to discuss what is working and what is not, and to build models for better practice. A balance needs to be struck between discussing and interpreting experiences and analyzing the scholarly works of others. Articles and guest speakers are common resources for the learning of group members.

Teachers in a professional practice school typically adopt a spirit of inquiry by asking questions about the curriculum, teaching, and learning. In seeking answers to these questions, teachers become learners in their own classrooms and use research to improve practice. Lieberman and Miller (1990) emphasized the importance of providing the conditions necessary for teachers to develop the knowledge and confidence to undertake systematic research. In addition to those conditions, the main resources needed in a professional practice school are time and commitment from administrators and colleagues to support an expanded teaching role.

The content of the curricula of a professional practice school is negotiated among the school's students, teachers, and school administrators. Lieberman and Miller (1990) use the term "content-in-context" to acknowledge the complexity of teaching and learning. This term provides scope for flexibility and diversity and, at the same time, manages to maintain the legitimacy of the content areas and the teacher's responsibility to teach something of value. Content coverage becomes less of a driving concern while student learning becomes the criterion in making curricular decisions.

Teachers in a professional practice school usually believe that the curriculum ought to serve the learning needs of students and be adaptable to a changing set of conditions that apply in the school. Having adopted new roles as curriculum designers, researchers, and mentors, the accountability structures that apply within a professional practice school might be expected to differ from what is applicable in schools in which traditional teaching and learning roles prevail.

ASSESSMENT OF TEACHER PERFORMANCE

Calls for accountability in education are increasing. It is apparent that increased spending does not translate into increased achievement among learners. Accordingly, there are renewed calls to assess teacher performance at a time when teachers are challenging the validity of the assessment systems that have been implemented. This trend alone would provide justification for formulating alternative approaches to teacher assessment. However, as teaching roles change, the criteria for assessment should be reconsidered and, as new ways of conceptualizing teaching emerge, it is possible to construct alternative models for teacher assessment.

Performance assessment systems based mainly on classroom observations have been built on the results of objectivist-oriented research on teaching and learning. Teachers from all grade levels and all subject areas are required to demonstrate given performance indicators on demand in a manner that is analogous to a person seeking a driver's license. The assumption is that competent teachers can demonstrate each of the performance indicators when asked to do so. What is observed, then, is a lesson that is planned to demonstrate what can be done rather than what typically is done. Teachers know the criteria for assessment and know when they are to be assessed. In most instances, they can choose from among their classes the one that they will use as a basis for the assessment. However, the approach is basically reductionist, with teaching often being divided into more than 100 performance indicators. At issue in assessment systems based on such approaches is whether or not such a reductionist view of teaching is viable. If one sums the set of performance indicators, does the sum represent teacher performance in a meaningful way? Is a person who scores satisfactorily on all or nearly all of the performance indicators guaranteed to be an effective teacher? And is a person who fails to score at a satisfactory level on one or many of the indicators likely to be unsuccessful? An additional point challenges the validity of such assessment schemes from the standpoint that generic performance indicators are problematic. Is it possible that any criterion can be applicable across all the grades from kindergarten through to Grade 12 and across subject areas?

The content of instruments for assessing teacher performance typically has been derived from research on teaching and learning.

Researchers carefully place limits on their studies and describe those limitations. In most instances, the size of the relationships between teaching performance and student achievement is modest. However, when the findings are applied in teacher performance instruments, bivariate relationships are assumed and the caveats surrounding relationships are absent. As performance indicators, research findings become a set of truths to which teaching is referenced. In most instances, these truths are based on the epistemology of objectivism. For example, for many years the conventional wisdom of teachers has been to control student behavior so that the class is quiet. Indeed, research programs have been premised on this assumption. Accordingly, the research literature provides lists of teacher behaviors and strategies that have been validated in terms of achievement in classes in which it is assumed that teachers are controllers of students. Can life in classrooms be reduced to such a simple assumption? Are there circumstances in which it makes sense for students to assume control? Should students learn to assume control at some time in their lives at school? From a constructivist perspective, it can be argued that students ought to be managed with learning as a priority, rather than having control as a priority as is the custom in schools. Instead of endeavoring to maintain silence, a classroom might be managed so as to enable students to talk with one another and utilize collaborative and cooperative learning strategies in the process of negotiating meaning. Student self-management might be a higher priority than management from the teacher. If the assumption of teachers as controllers of students is abandoned, there is little research to guide teachers in the selection of practices that are likely to facilitate learning.

How might teacher performance be assessed? Wolf (1991, p. 5) noted that "any system of teacher assessment must faithfully reflect the richness and complexities of teaching and learning." A teacher performance assessment system based on a constructivist perspective would provide teachers with greater autonomy with respect to what is to be assessed, when assessments will occur, and the conditions under which an assessment will take place. There ought to be a clear link between assessment and what is known. A valid measure of a particular construct, then, would be the extent to which the measure represents knowledge of the construct. Is it possible to conceptualize assessment as an opportunity for teachers to demonstrate what they know? If such a definition were adopted, would it be possible to develop a teacher assessment system? If so, what would such a system look like? To begin with, teachers would have the freedom to choose when and how they would be assessed. Furthermore, they would have a degree of autonomy over the contextual factors associated with an assessment. For example, for a given set of competencies, teachers might decide to demonstrate their knowledge in a simulation situation. Or, teachers might opt for an assessment based entirely on performance in their classrooms. In the latter instance, it would make sense to allow the teacher to select the class or classes in which an assessment would be conducted, as well as to select the topics to be taught and used as the basis for an assessment. Teachers are given more control, and hence responsibility, in the assessment process. Of course, although giving the teacher greater responsibility will not guarantee validity of the measures of performance, it does seem to increase the probability that an assessment will provide teachers with opportunities to demonstrate their knowledge and its application to practice.

For the past twenty years, the most promising approaches to teacher assessment have been based on direct observation of teaching in conjunction with focused judgments on the effectiveness of specified aspects of teaching and learning. The pioneers of numerous teacher assessment instruments in the United States have been William Capie, from the University of Georgia, and Chad Ellett, from Louisiana State University. Together and separately, these individuals refined a technology of teacher assessment that takes account of validity, reliability, and credibility (Ellett, Loup, and Chauvin, 1991). As one state after another in the United States joined the move toward greater accountability of teachers, the content and structure of teacher assessment instruments improved. However, among the bulk of teachers, these instruments were not popular. They were perceived as threatening and unwieldy. To many teachers, the assessment systems lacked credibility. Accordingly, there has been a recent trend for the assessment systems that were developed and implemented in states such as Georgia and Louisiana to be discontinued. The political backlash against using assessment instruments that did not have the support of unions and the majority of teachers was too great.

The search for viable systems for the assessment of the performance of teachers, especially systems to link assessment with the professional growth of teachers, continues in the United States. A group from Stanford University, led by educators such as Lee Shulman and Angelo Collins, has explored different approaches to the assessment of teacher performance. These approaches avoid overly reductionist views of teaching and do not embrace the necessity of basing assessments almost entirely on in-class performance. The Stanford group has given considerable attention to *assessment centers* and *portfolios.*

An assessment center is a site for the assessment of teachers. As such, it is a place where teachers can demonstrate whether or not they have the knowledge needed to be an effective teacher. Because the center is remote from actual classrooms, teachers can engage in a variety of tasks that enable them to demonstrate what they know about curricula, teaching, and learning. For example, computer simulations of teaching and/or learning can be used to set a context for the application of professional knowledge, and portfolios can set the stage for personal reflections on classroom practices and deliberations on the efficacy of teaching practices.

Portfolios enable teachers to document their teaching in the authentic setting of their own classroom. Wolf (1991) noted that "when the actual artifacts of teaching are combined with the teacher's reflections, portfolios permit us to look beneath the surface of the performance itself and examine the rationales and decisions that shaped the teacher's actions" (p. 23). Shulman (1988, p. 40) suggested that portfolios "retain almost uniquely the potential for documenting the unfolding of both teaching and learning over time and combining that documentation with opportunities for teachers to engage in the analysis of what they and their students have done." Portfolios provide teachers with the power to decide what best represents their knowledge. If a portfolio is used as a part of an assessment center's approach to teaching, teachers can include both artifacts of teaching and learning and written reflections on the significance of these artifacts and classroom activities. Included in the reflective comments are justifications for including the artifacts as evidence of what is known. Teachers can attach brief written captions which identify and explain the purpose of each piece of evidence in the portfolio. Videotapes of performance also can be included as evidence of teacher knowledge, thereby providing opportunities to evaluate both an actual teaching episode and, through reflective statements and a follow-up interview at the assessment center, the teacher's own assessment of that same event can be obtained. The contents of the portfolio provide evidence of the methods and materials used by a teacher in his or her classroom and the reflective commentaries allow the evidence to be examined in the light of the teacher's goals and beliefs.

Shulman described attempts to measure the performance of teachers as a "union of insufficiencies," meaning that no combination of methods to obtain data would prove to be sufficient to provide valid measures of teacher performance. The main implication of Shulman's position is that, even when efforts to measure teacher performance incorporate a range of approaches, the aggregated measures used to assess teaching performance are fallible. Accordingly, the use of portfolios alone will not ensure viable measures of teacher performance and it is recommended that assessment centers utilize a variety of approaches to obtain performance profiles for teachers seeking an assessment. There is a greater likelihood that an assessment will provide viable measures of what teachers know if teachers are provided with autonomy to suggest assessment activities and, in conjunction with center personnel, negotiate a comprehensive package of assessment activities such as portfolios, computer simulation activities, and reflective discussions.

IMPLICATIONS

Given the caveat that knowledge derived from research on teacher learning and change is not generically applicable, it is useful to ask what is known about teacher change and performance assessment after more than two decades of research. Is our cup of knowledge overflowing, half-full, or empty? In answer to this question, I suggest that the cup is well filled, but nowhere near the top. We know a great deal to guide policymakers and practitioners in their professional decisions, but there is much to be learned.

In this chapter I have considered research in several areas, including teacher education (strategy analysis, coaching, and learning as a basis for teacher change, teachers' metaphors), cultural perspectives on school change (constraints, myths), professional practice schools, and the assessment of teacher performance (assessment centers, portfolios). Although specific knowledge is not necessarily generalizable from one situation to another, the following implications are derived from the research reviewed in earlier sections of the chapter:

1. Teachers should study a teaching model, use it to plan lessons, analyze lessons using the model, and receive feedback from others about the extent to which teaching and learning processes conform to the model.

2. Teachers should observe one another's classes and discuss what did and did not work.
3. Teachers should conceptualize new teaching roles in terms of metaphors, analyze the curriculum in terms of metaphors embedded in the practices of teachers and students, and identify the situations in which specific metaphors are and are not appropriate.
4. Teachers should analyze curricular events in terms of the embedded beliefs about power relationships between the teacher and students and beliefs about the nature of knowledge, learning, and knowing.
5. Teachers should identify the constraints that prevent them from implementing the curriculum as they would prefer, identify those who have most control over each constraint, and plan to overcome the constraints.
6. In order to sustain changes successfully in a school, teachers should construct a personalized vision of the changed curriculum, construct a commitment to personal change, and reflect on professional actions.
7. Professional practice schools should be developed as collaborative partnerships between schools, universities, and the community to facilitate the learning of school staff and students, prospective teachers, and university staff.
8. Portfolios and simulations should be used in the assessment of teacher performance to ensure that teachers are able to demonstrate what they know and apply it in contexts which they consider to be relevant.

CONCLUSION

The decade of the 1990s offers hope for a new era in education. Fresh insights have arisen from the use of constructivism to make sense of teaching, learning, curriculum, and teacher assessment. The manner in which teacher knowledge is viewed places primary responsibility upon teachers to structure their work environments as learning environments and construct themselves as learners with respect to their work as teachers. No longer can teachers be assumed to be passive recipients of knowledge produced by university scholars and published in journals to be consumed by knowledge-hungry teachers. Instead, teachers are seen as researchers in their own classrooms and co-learners with their students in tackling research on learning, curriculum change, and cultural change on a daily basis. At the same time, teachers are the architects of curriculum reform. Curricula no longer can be thought of as entities produced by outside experts to be transported into classrooms around the world and implemented with fidelity. Instead, a curriculum becomes a personal construct for learners, defined by a unique culture which is a characteristic of the particular school and class in which it is implemented. With an expanded set of responsibilities, teachers face daunting challenges as educators of tomorrow's citizens.

Teachers ought not face the challenges of education alone. The idea of professional practice schools is appealing in that recognition is given to the need for teachers to be active constructors of their own knowledge in the social settings in which the knowledge has situated relevance. Knowledge is neither universal nor automatically generalizable from one school building to another. Furthermore, there is no guarantee that knowledge constructed one day can be applied to the same class one day later. Having teachers learn in their own professional practice communities, receive support from their colleagues, and apply what they learn for the immediate benefit of their students seem to provide a sensible, albeit different, concept of teacher education. Can the resources needed to transfer teacher education to professional practice schools be allocated for this purpose? Clearly, such a redistribution of resources has political implications because teacher educators also would have to construct new roles in relation to the education of prospective and practicing teachers. Similarly, if teachers are to assume responsibility for the learning of their students, they have to be empowered to assume control and be accountable for their professional actions. Accordingly, a new generation of teacher assessment systems will be needed to allow teachers to demonstrate their competence in an expanded arena in which they have assumed new roles and have implemented curricula in ways that possibly are radically different than traditional approaches.

The most promising approach to the accountability of teachers is to use a variety of strategies to assess what teachers know and the extent to which their knowledge assists student learning. There is no doubt that the obvious face validity of assessing teachers as they teach will lead to the development of a new generation of instruments for teacher assessment that focus on classroom performance and take account of the expanded roles of teachers. Such instruments more than likely will provide teachers with increased autonomy with respect to the criteria for an assessment and the conditions under which assessments occur. The notion of assessment centers also is appealing. Such centers provide teachers with opportunities to demonstrate what they know in meaningful contexts that simulate classrooms and curricula that have personal relevance. Although the most appropriate tasks to be used at an assessment center are yet to be determined, the use of portfolios has considerable untapped potential. A portfolio provides teachers with the autonomy to show what they know and what they have learned over a period of time. However, the portfolio is not seen merely as an end in itself, but as a focus for reflective discussions with assessment center staff and as a context for teachers to represent what they know about teaching.

Thinking about teacher change in terms of teacher learning, curriculum change, and cultural transformation leads to new approaches to teacher education, expanded roles for teachers in schools, and different perspectives on accountability. In this chapter I have reviewed the impressive growth in our understandings of teaching and learning over the past twenty years, and I have also anticipated the excitement of the next twenty years. Can the educational systems of the world restructure so as to take advantage of our knowledge of how teachers learn, how curricula can be reformed, and how school cultures can be changed to facilitate learning? Inevitably the issue of change rests with individuals, but all individuals act within social settings in which cultural change occurs slowly as differences between individuals are resolved through negotiation and consensus building. The possibility of change can be increased if individuals within the culture come forward as champions of change. Who will be those champions? Unless they emerge and speak with clear voices, the status quo will continue in the form of traditional approaches to teacher education, curriculum reform, and tried and tested approaches to fixing an educational system which is teetering on a crumbling base of objectivism.

REFERENCES

Belenky, Mary F.; Clinchy, Blythe M.; Goldberger, Nancy R.; and Tarule, Jill M. *Women's Ways of Knowing: The Development of Self, Voice, and Mind.* New York: Basic Books, 1986.

Britzman, Deborah. *Practice Makes Practice: A Critical Study of Learning to Teach.* Albany, N.Y.: SUNY Press, 1991.

Ellett, Chad D.; Loup, Karen S.; and Chauvin, Sheila W. "Development, Validity and Reliability of a New Generation of Assessments of Effective Teaching and Learning: Future Directions for the Study of Learning Environments," *Journal of Classroom Interaction* 26, no. 2 (1991): 25-39.

Fien, John. "Ideology, Political Education and Teacher Education: Matching Paradigms and Models," *Journal of Curriculum Studies* 23 (1991): 239-256.

Grundy, Shirley. *The Curriculum: Product or Praxis?* London, England: Falmer Press, 1987.

Hart, E. P., and Robottom, Ian M. "The Science-Technology-Society Movement in Science Education: A Critique of the Reform Process," *Journal of Research in Science Teaching* 27 (1990): 575-588.

Joyce, Bruce R., and Showers, Beverly. *Power in Staff Development through Research on Training.* Washington, D.C.: Association for Supervision and Curriculum Development, 1983.

Lambert, Linda. "Staff Development Redesigned," *Phi Delta Kappan* 69 (1988): 666.

Lieberman, Ann, and Miller, Lynne. "Teacher Development in Professional Practice Schools," *Teachers College Record* 92 (1990): 105-122.

Little, Judith W. "District Policy Choices and Teachers' Professional Development Opportunities," *Educational Evaluation and Policy Analysis* 11 (1989): 165-179.

Munby, Hugh, and Russell, Thomas. "Metaphor as an Instructional Tool in Encouraging Student Teacher Reflection," *Theory Into Practice* 29 (1990): 116-121.

Russell, Thomas. "Learning to Teach Science: Constructivism, Reflection, and Learning from Experience." In *Constructivist Perspectives on Science and Mathematics Education*, edited by Kenneth Tobin. Washington, D.C.: American Association for the Advancement of Science, in press.

Sanders, Donald P., and McCutcheon, Gail. "The Development of Practical Theories of Teaching," *Journal of Curriculum and Supervision* 2 (1986): 50-67.

Sarason, Seymour B. *The Culture of the School and the Problem of Change*, 2d ed. Newton, Mass.: Allyn & Bacon, 1982.

Shulman, Lee. "A Union of Insufficiencies: Strategies for Teacher Assessment in a Period of Educational Reform," *Educational Leadership* 46, no. 3 (1988): 36-41.

Tobin, Kenneth; Davis, Nancy T.; Shaw, Kenneth L.; and Jakubowski, Elizabeth H. "Enhancing Science and Mathematics Teaching," *Journal of Science Teacher Education* 2, no. 4 (1991): 85-89.

Tobin, Kenneth, and Espinet, Mariona. "Teachers Helping Teachers to Improve High School Mathematics Teaching," *School Science and Mathematics* 90 (1990): 232-244.

Tobin, Kenneth; Kahle, Jane B.; and Fraser, Barry J., eds. *Windows into Science Classrooms: Problems Associated with Higher Level Cognitive Learning.* London, England: Falmer Press, 1990.

Tobin, Kenneth, and Ulerick, Sarah J. "An Interpretation of High School Science Teaching Based on Metaphors and Beliefs for Specific Roles." Paper presented at the Annual Meeting of the American Educational Research Association, San Francisco, 1989.

von Glasersfeld, Ernst. "Cognition, Construction of Knowledge, and Teaching," *Synthese* 80 (1989): 121-140.

Wolf, Kenneth. "The Schoolteacher's Portfolio: Issues in Design, Implementation, and Evaluation," *Phi Delta Kappan* 73 (1991): 129-136.

Yeany, Russell H. "The Effects of Model Viewing with Systematic Analysis on the Science Teaching Styles of Preservice Teachers," *Journal of Research in Science Teaching* 14 (1977): 209-222.

Yeany, Russell H., and Porter, Charles F. "The Effects of Strategy Analysis on Science Teacher Behaviors: A Meta-Analysis." A paper presented at the Annual Meeting of the National Association for Research in Science Teaching, Lake Geneva, Wis., 1982.

Chapter 8

USE OF COMPUTERS

Tjeerd Plomp and Joke Voogt

New technologies are transforming our society into an information society. Information technology plays an important role in every corner of the Western world, in our daily lives, and in the development and application of science. Science education should contribute to preparing students to live in a world where the new technologies are so widespread. Furthermore, the use of educational technologies should contribute to improving teaching and learning in science.

In this chapter, we deal mainly with the use of computers in science education, focusing on the ways in which they are used and on evidence of their effectiveness. We report an international survey that illuminates the extent of availability and use of computer hardware and software in schools, the changes in classroom instruction that teachers attribute to the use of computers, the problems which teachers say that they encounter in using computers, and the reasons which teachers give for not using computers. Finally, we make suggestions for minimizing the problems of implementing the use of computers and for achieving effective integration of computers into science teaching. Although the chapter deals mainly with the use of computers in science education, some remarks are made about other new technologies, such as interactive video technology.

APPROACHES TO THE TEACHING OF SCIENCE

In the past thirty years, various approaches to science education have been developed. As Roth (1989) observes, however, most science

Chapter consultants: Karl Frey (University of Zurich, Switzerland) and Pavla Zieleniecova (Charles University, Czechoslovakia).

classes still are dominated by programs that emphasize scientific facts and theories. The alternative approaches have not been implemented widely in classrooms. Roth describes three important approaches to science education that are found particularly in English-speaking countries and some other countries such as The Netherlands. Although Roth's study relates to elementary schools, it applies also to secondary education (Millar, 1985). The three approaches are described briefly below.

The inquiry approach. In the 1960s and 1970s, new science programs were developed to emphasize the nature and processes of science. In these programs, students were to act as scientists, and the scientific method was the most important part of the curriculum. Science content served as a means for learning science process skills. From meta-analyses, we know that these programs generally have had positive effects on students' achievement and motivation (Shymansky, Kyle, and Alport, 1983).

The societal approach. In the late 1970s and 1980s, programs were developed to emphasize science as part of our technological society. Central concerns in these programs were the teaching and learning of skills involved in decision making and problem solving in relation to societal problems. Students were expected to become responsible citizens and the content of science education was derived from societal issues.

The conceptual change approach. From the mid-1980s, programs were developed to stress conceptual change in learning science. Their starting point is the idea that the learner constructs a world view based on personal observation and experience (Linn, 1987). The content of science is derived from key science concepts.

In contrast to science programs which stress facts and theories, which we call the "traditional approach," the three new approaches above involve learning environments that assume an active role for students. Why have these new approaches not been implemented widely in science classrooms? It appears that problems of time, the changing role of the teacher, and the difficulty of designing rich learning contexts impede the implementation of these approaches (Roth, 1989). We outline here some of the problems of implementation, as well as the potential of new technologies in science teaching.

In the conceptual change approach, the learner actively constructs his or her own framework (see chapter 3 in this volume). Starting points in science education are the children's prior conceptions. Instructional activities should be designed to make children's reasoning explicit and to confront them with the consequences of their reasoning. This approach is time consuming and presupposes the individualization of instruction. New technologies, such as computer simulations, can help to make the reasoning of children explicit and help them to visualize the consequences of their thinking as they work individually or in small-group settings (Driver and Scanlon, 1988).

Especially in the inquiry approach, but also in the other two approaches as well, laboratory work takes a central position in science lessons. Tobin (in Shymansky and Kyle, 1988) noted that most laboratory activities in science lessons are of a cookbook type with most emphasis being directed to collecting data. Few opportunities are provided for students to plan investigations or to interpret data. With new technologies, such as microcomputer-based laboratories, data collection and graphic representation of data can be handled by the computer, thus allowing students to focus more on the design of experiments and on the interpretation of the data.

The starting points of the societal approach are current issues confronting our society. But often it is difficult to design for students interesting and rich contexts in which they can experiment with data in an attempt to get a grasp of the issues. First, a number of experiments cannot be conducted for safety and/or ethical reasons. Second, it is often difficult to show the long-term effects of technological applications in society. Third, it is difficult to provide students with enough empirical data. These problems can be overcome in part with new technologies such as databases, computer simulations, and interactive video technology.

New technologies are useful not only for changing the method of science education; new technologies themselves also can be a part of the content of the curriculum. This is advocated especially in the societal approach to science education.

All three approaches call for a change in the role of the teacher. Instead of presenting scientific facts and theories to be learned, teachers become facilitators of the learning process. An active learning environment demands planning and management skills from teachers. A lack of these skills can be a barrier to implementation (Roth, 1989; Shymansky and Kyle, 1988). In facilitating learning, teachers can use new technologies, such as computer-managed instruction, as tools.

POTENTIAL OF COMPUTERS IN SCIENCE EDUCATION

Before analyzing the potential of some of the new technologies in science education, we describe their essential characteristics and then discuss their anticipated effects in science education. Lauterbach and Frey (1987) classified computer programs, including those that structure students' work and those in which students themselves provide the structure for their own work and learning:

1. *Computer-managed instruction.* These programs take over classroom management functions such as testing, feedback on test results, and directing students to other learning activities.

2. *Computer-assisted instruction.* Many different programs, from simple "drill-and-practice" programs to complex tutorials, belong to this type of application. Drill-and-practice programs are meant for repetition and consolidation of skills. No new information is given. Tutorials guide the student's learning path. Computer-assisted instruction entails the learning of new information. Individualizing the learning process becomes possible.

3. *Simulations, modeling systems, and microworlds.* Simulations are a representation of a part of reality. Students can gain understanding of the reality by manipulating this representation. Through simulation, it is possible to study parts of reality which could not be studied before for reasons of safety, ethics, cost, lack of apparatus, or scale. Simulations can be an aid in visualizing abstract concepts as well as a bridge between reality and the student's mental model of the system (van Schaick Zillesen, 1990).

Modeling systems and microworlds can be considered specific types of simulation. With dynamic modeling systems, the students themselves build their own model of a part of reality. They gain understanding of complex relations in a system. Microworlds do not aim to represent reality. They are imaginary worlds in which it is possible for students to investigate scientific problems, develop hypotheses, design experiments to test their ideas, and use feedback to reflect on their conceptions of the phenomena (Linn, 1988).

4. *Microcomputer-based laboratories*. With microcomputer-based laboratories, data collection is possible. It becomes much easier to repeat experiments, measure different variables at the same time, use a very short or long time range, analyze data, and represent data graphically. Instead of spending valuable instructional time in data collection, students can use microcomputer-based laboratories for analyzing and interpreting data. For microcomputer-based laboratories, supplementary hardware is necessary.

5. *Databases*. Information from various areas of knowledge is stored in databases and can be retrieved readily. Having retrieved information, students then can use it in practicing problem-solving skills.

6. *Tools*. General software tools can be used in science education as well. For example, spreadsheets can be used for calculations and word processing can be used for writing reports of laboratory experiments.

7. *Interactive video technology*. This technology can be considered as an extension of the computer applications already mentioned, especially tutorials. A videodisc can provide realistic pictures instead of written text or a graphic representation.

While computer applications are believed to have much potential for science education, the anticipated effects of using new technologies often are not yet supported by empirical evidence.

It often is asserted that new technologies enhance students' performance and motivation. For science education specifically, simulations, microcomputer-based laboratories, and databases are said to be important for facilitating mastery of science concepts and of science process skills.

From several meta-analyses (Bangert-Drowns, Kulik, and Kulik, 1985; Roblyer, Castine, and King, 1988) and syntheses of meta-analyses (Frey, 1988; Niemiec and Walberg, 1987), we know that computer-assisted instruction is most successful in special education and elementary school education. In secondary education, the effect of computer-assisted instruction on student performance is relatively small. As students grow older, the effect of computer-managed instruction increases. The effect of computer applications intended to enhance the curriculum (simulations, databases, and microcomputer-based laboratories) were not known very well until quite recently. Roblyer et al. (1988)

found that science simulations had a positive effect on performance for college and senior high school students. In other reviews (Bangert-Drowns, Kulik, and Kulik, 1985; Frey, 1988; Niemiec and Walberg, 1987), simulations were not found to have a substantial effect on students' performance.

These reviews, however, were based on rather old studies. The effects of the use of databases or microcomputer-based laboratories on students' performance have not been included in meta-analyses until recently. Several studies (Linn and Songer, 1991; Mokros and Tinker, 1987) showed that students' graphing skills increased when microcomputer-based laboratories were used. Research on the mastery of science process skills or science concepts through using simulations or microcomputer-based laboratories still is going on. Until now, the results show that students working with simulations are doing at least as well as control groups (Dekkers and Donatti, in Frey, 1988).

The results from these meta-analyses show that working with computer applications is motivating for students and that their attitudes toward computers and toward instruction improve with the use of computer applications.

One well-known potential associated with using new technologies in education is the possibility of realizing individualized instruction. Suppes and Fortune (1985) mention five ways of individualizing instruction, each of which is helped by computer-assisted methods. The interactive videodisc, with its capability of easily providing random access and a large range of presentation modes, also might help in individualizing instruction (Bijlstra, 1986).

Because of a lack of computers in schools, most computer work done by students takes place in small-group settings involving at least two students. One unanticipated finding in Linn's study of the use of microcomputer-based laboratories was that cooperation between students was promoted (Linn and Songer, 1991; Striley Stein, 1986-1987). Students controlled their results through comparing them with others. Hawkins and Sheingold (1986) and Cox (1992) also reported that collaborative learning occurred during the use of databases and simulations. In her study of microcomputer-based laboratories, Striley Stein found an extremely low incidence and short duration of off-task periods during lessons. In another study of microcomputer-based laboratories, Findhammer et al. (1986) found the same results. Frey (1988) concluded that time-on-task should be an explicit variable in studies assessing the effects of using new technologies in education.

Several authors (Roblyer et al., 1988) consider an increase of approximately 10 percent in time for learning as an important gain when using computer applications such as computer-assisted instruction. For interactive videodisc technology, a similar gain in learning time has been reported (Cushall et al., 1987). An important aspect of computer applications is the release of the student from tedious labor (Cox, 1992). As a result, more time can be spent on other parts of the curriculum or on the learning of science concepts or process skills.

Research on computer applications that aim to enhance the curriculum is not very promising. Linn and Songer (1991) argue that the full potential of microcomputer-based laboratories for learning science concepts and science process skills can be realized only in a science curriculum that treats a few topics thoroughly instead of many topics superficially. This probably also is true for the use of simulations and databases.

While the new technologies provide rich potential for science education, their use implies significant changes in the daily practices in science classrooms and laboratories. Probably the content needs to be changed so that fewer topics will be covered but each topic will be considered in greater depth than is possible when a wide range of topics is included in the curriculum (Linn, 1988). Based on their studies of the integration of computers into the learning process, Hawkins and Sheingold (1986) concluded that teacher-student and student-student interactions change qualitatively. Teachers are not necessarily always "at center stage." They need time to organize effective collaboration among students, and they need skills in observing and intervening in the group work of students. These kinds of management skills are different and more complex than those used in traditional science classes.

STATUS OF COMPUTER USE IN SCIENCE EDUCATION

The International Association for the Evaluation of Educational Achievement (IEA) conducted an international comparative survey (Computers in Education [Comped]) in which data were collected for school and teacher variables in nineteen countries during 1989 (Pelgrum and Plomp, 1991a). The populations of interest for this study were located in elementary, junior secondary, and senior secondary education. Data were collected in schools that used computers as well as in schools that did not use computers, and from computer-using teachers as well as teachers who were not using computers. In this section, we briefly describe the procedures for collecting the data and summarize some of the data from science teachers in junior and senior secondary education.

The samples of teachers in junior and senior secondary education included not only computer education teachers, but also teachers of mathematics, science, and the mother tongue. The relevant populations were defined as follows:

● *Non-using schools.* All schools which did not use computers for teaching/learning purposes in grades in which the modal age of students was 12, 13, or 14 years (junior secondary level) and in the final or penultimate secondary grades (senior secondary level).

● *Using schools.* All schools in which computers were used for teaching/learning purposes in grades in which the modal age of students was 12, 13, or 14 years (junior secondary level) and in the final or penultimate secondary grades (senior secondary level).

● *Computer-using science teachers.* All science teachers in computer-using schools who provided lessons in subjects in which computers were used in grades in which the modal age of student was 13 years (junior secondary level) and by students who were in their final year of secondary education (senior secondary level).

● *Non-computer using science teachers.* All science teachers in computer-using schools who provided lessons in which computers were used neither in grades in which the modal age of students was 13 years (junior secondary level) nor by students who were in their final year of secondary education (senior secondary level).

Data were collected by means of questionnaires that were answered by principals, computer coordinators, and teachers. Pelgrum and Plomp (1991a) provide complete and detailed information about the design and results of this survey.

The following generalizations from the Pelgrum and Plomp report provide a broad overview of the nature and extent of computer use in the sample of junior and senior secondary schools in the nineteen countries included in the study:

1. With but few exceptions, computers were being used for instructional purposes in a majority of the junior and senior secondary schools in the sample.
2. The median number of computers per school in the sample ranged from 2 to 43 at the senior secondary level and from 5 to 43 at the junior secondary level.
3. In most schools, computers were located in special computer classrooms.
4. The most intensive use of computers for instructional purposes was by computer education teachers.
5. In most countries, only a limited number of science teachers were using computers for instructional purposes.

When computer-using science teachers were asked about the approaches they take when using computers in their science lessons, we found that the most common practice among the responding teachers was to have students use computers in a "drill mode" (that is, the students do exercises with the computer). Computer-assisted instruction is quite common, as is the use of the computer by the teacher in demonstrations and explanations. We were surprised to find that computers are used infrequently for the purposes of remediation.

Computer-using science teachers also were asked about changes that occurred in their classes as a consequence of computer use. The following changes were mentioned frequently: increased student interest in the subject matter; increased cooperation among students (students tutoring and helping each other, students working in small groups); improved achievement of students in science; an increase in the amount of science curriculum that was "covered" (especially in the senior secondary school). But a substantial percentage of the teachers reported that it was more difficult for them to organize lessons and that they needed more time for preparing lessons when computers were to be used.

Among the problems computer-using science teachers encountered when introducing computers into the instructional program were an insufficient number of computers and a lack of suitable software, difficulties in integrating computer use into the instructional program, and difficulties in scheduling enough computer time for their classes. When science teachers who were not using computers were asked why they were not involved in computer use, many of them cited the problems mentioned by computer-using science teachers. But non-users also mentioned (more frequently than the users) their lack of knowledge and skills relating to computers.

INTEGRATING COMPUTERS INTO THE SCIENCE CURRICULUM

From the data obtained in the Comped study, and especially from the data obtained from computer-using science teachers and from science teachers not using computers, we see a picture that can be interpreted in different ways. If the present situation regarding computer use in many schools is compared with the potential of computers for the science curriculum, then our conclusion must be that possibilities are far from being realized. Efforts of national, state, regional, local, and school authorities apparently have resulted in only limited use of computers in science education. Only a small number of science teachers are using computers for instructional purposes. Furthermore, the types of use are not very advanced. The pattern that we found is not much different from Walker's (1986) conclusion about developments in the USA in 1983-1984. The results reflect an implementation process in an early stage of development. According to Walker, the easiest way for schools to respond to the challenge to "join the computer revolution" is to start with the easiest applications such as drill-and-practice activities that involve taking the whole class to a computer laboratory. Walker points out that "anything else requires more money, more effort and expertise from teachers, and more variance from existing school practices" (p. 35).

We do not think that we should be disappointed by this situation, particularly if one regards the use of computers in education as a complex innovation. Walker (1986, p. 33) states: "If even a small part of the visionary dreams of computer-based education is to be realized, major changes will be required in the day-to-day activity and interaction patterns in classrooms. . . . Developing these new patterns will require collaborative effort on a large scale sustained over a decade or more." From this perspective, the present situation with regard to computer use in education can be considered as the beginning stage of a process that could last many years. If administrators, teachers, and courseware developers consider the present situation from such a perspective, and if they are willing to undertake initiatives consistent with that situation, then we can expect a development in science education away from the easiest responses to the technological challenges.

From their discussion of strategies for implementing microcomputers in schools, Fullan, Miles, and Anderson (1988) say that a precise end-state of "good implementation" in terms of effective teaching strategies and of effective use of specific software is uncertain for this innovation. They argue that the scale of the innovation is large and the time line long, that it is unclear how to accommodate the old and new, and that there are wide variations in teacher commitment and experience with different types of hardware and software. They suggest a process planning or adaptive approach in which a variety of strategic initiatives can be launched and in which there is a strong emphasis on learning from practical experience. They believe that the most promising strategies should center around developing the competence of teachers, training consultants, diffusing and supporting effective practices, networking, and building organizational capacity. Accordingly, in order to reduce the problems which science teachers encounter in implementing the use of computers in their teaching, practices such as the following are likely to be helpful:

1. For non-users, the emphasis should be on the development of knowledge and skills related to computers. For users, training should emphasize the integration of software into instruction. Thijsen et al. (1990) recommended that an effective approach for teachers who are beginning to use computers is to train them within the context of their schools by convincing examples of computer applications and providing them with ample hands-on experience.

2. Training of local or regional consultants who are able to support teachers in identifying and selecting software, to stimulate grass-root initiatives such as experiments, pilot studies, and demonstrations, and to disseminate and support the implementation of well-developed practices.

3. Developing networks so that using teachers within and across schools can share approaches to stimulate the interest of other teachers.

4. Building organizational capacity so that school boards and administrators are able to manage the change process.

Fullan et al. (1988) and Pelgrum and Plomp (1991b) propose some short-term strategies to promote the integration of computers into the classroom. They suggest the use of computers by teachers in demonstrations as a short-term strategy to accommodate the limited number of computers in schools. While this can be very appropriate for some laboratory experiments and simulations, there is the disadvantage that students cannot profit from interacting directly with a computer. Furthermore, use of the computer in whole-class activity has consequences for the software to be acquired.

Among the important problems seen by teachers in using computers are the lack of time for preparing lessons and the lack of computer-related skills. Pelgrum and Plomp (1991b) suggest that these problems can be counteracted in part by improving the quality of the software. Van den Akker, Keursten, and Plomp (1992) reviewed the literature on problems associated with the quality of educational software and courseware. Their starting point is a curriculum implementation perspective which emphasizes the role of the teacher. They point to the following problems:

1. Much software contains material that is attuned badly to the curriculum.
2. Much software has been developed for use by individual students, without taking into account the more usual use of computers in whole-class settings and the fact that most classrooms have only a limited number of computers.
3. Much software does not include support for the teacher on how to integrate the use of the software with other instructional activities.
4. Much software is still of drill-and-practice or tutorial type and does not exploit sufficiently the capacity of the computer for more advanced uses to enhance teaching and learning.
5. Most software is just piecemeal and deals with only a limited part of the subject matter about which students are to be informed.

The lack of high-quality materials makes teachers feel uncomfortable about selecting suitable software, about their skills when using the computer, about unanticipated problems in a new kind of instruction, and about managing and organizing activities that emphasize computer use.

Van den Akker et al. (1992) note that the integration of computers into teaching requires that teachers learn new methods of instruction and this involves changed beliefs about how students learn, changed attitudes toward the use of technology in the classroom, and a deepened understanding of the potential of computers for instruction.

Voogt (1990) studied teachers' evaluations of software for science education. She concluded that teachers should make their own cost-benefit analysis before using software in the science curriculum. The benefits of using software in science lessons are obvious when the software (a) motivates students, (b) enables students to achieve educational objectives, and (c) has content which operationalizes the teacher's beliefs and ideas, so that the software can narrow the gap between beliefs and practice in science education. To achieve these conditions, constant attention should be given to formative evaluation and revision of software.

RECOMMENDATIONS

In conclusion, we suggest recommendations for the introduction of computers into science education, as well as for the further improvement of courseware for use with computers:

1. The introduction of computers should take into account the specific characteristics of this new technology and the amount of time likely to be required to introduce them effectively in the classroom.

2. Introducing computers should be accompanied by a support structure that incorporates the development of teachers' competence in the use of computers, the training of consultants, the provision of network facilities, and the building of organizational capacity.

3. Teachers inexperienced in the use of computers in the classroom should receive in-school training in which they can observe good examples of computer use and have much hands-on experience with computers.

4. Measures should be taken to insure that the courseware developed outside the schools and selected by schools supports the existing goals of the curriculum or will be helpful in securing desired changes in the curriculum.

5. Courseware developed for school use should be designed carefully and include "how to do it" advice to support its appropriate use by teachers.

6. Much attention should be given to formative evaluation of courseware in classroom settings as a basis for making needed revisions and improvements in the materials.

REFERENCES

Bangert-Drowns, Robert L.; Kulik, James A.; and Kulik, Chen-Lin C. "Effectiveness of Computer Based Instruction in Secondary Schools," *Journal of Computer-based Instruction* 12 (1985): 59-68.

Bijlstra, Jim. *Interactieve Video En De Opleiding Van Operators: Een Literatuurstudie* [Interactive Video and the Training of Operators: A Literature Review]. Enschede, The Netherlands: Department of Education, University of Twente, 1986.

Cox, Margaret J. "The Computer in the Science Curriculum," *International Journal of Educational Research* 17 (1992): 19-35.

Cushall, Marcia B.; Harvey, Francis A.; and Brovey, Andrew J. "Research on Learning from Interactive Videodiscs: A Review of the Literature and Suggestions for Future Research Activities." Paper presented at the meeting for Educational Communications and Technology, Atlanta, GA, 1987.

Driver, Rosalind, and Scanlon, Eileen. "Conceptual Change in Science," *Journal of Computer Assisted Instruction* 5 (1988): 25-36.

Findhammer, Wim J.; Verkerk, Gerrit; and Heijeler, Ronald. "Het Gebruik Van Een Micro-computer Bij Het Practicum Natuurkunde" [The Use of a Microcomputer for Laboratory Work in Physics], *Tijdschrift Didactiek B-wetenschappen* 4 (1986): 168-177.

Frey, Karl. *Effectiveness of Computers in Education: A Survey of Empirical and Meta-Analytical Studies*. Zurich, Switzerland: Institut für Verhaltenswissenschaft, Eidgenoissische Technische Hochschule, 1988.

Fullan, Michael G.; Miles, Matthew B.; and Anderson, Stephen A. *A Conceptual Plan for Implementing the New Information Technology in Ontario Schools*. Ontario, Canada: Ministry of Education, 1988.

Hawkins, Jan, and Sheingold, Karen. "The Beginning of a Story: Computers and the Organization of Learning in Classrooms." In *Microcomputers and Education*, Eighty-fifth Yearbook of the National Society for the Study of Education, edited by Jack A. Culbertson and Luvern L. Cunningham, pp. 40-58. Chicago, University of Chicago Press, 1986.

Lauterbach, Roland, and Frey, Karl. "Educational Software: Review and Outlook," *Prospects* 17 (1987): 387-395.

Linn, Marcia C. "Establishing a Research Base for Science Education: Challenges, Trends and Recommendations," *Journal of Research in Science Teaching* 24 (1987): 191-216.

Linn, Marcia C. "Science Education and the Challenge of Technology." Paper presented at the Annual Meeting of the American Educational Research Association, New Orleans, LA, 1988.

Linn, Marcia C., and Songer, Nancy B. "Teaching Thermodynamics to Middle School Students: What Are Appropriate Cognitive Demands?," *Journal of Research in Science Teaching* 28 (1991): 885-918.

Millar, Robin. "Training the Mind: Continuity and Change in the Rhetoric of School Science," *Journal of Curriculum Studies* 17 (1985): 369-382.

Mokros, Janice R., and Tinker, Robert F. "The Impact of Microcomputer-Based Labs on Children's Ability to Interpret Graphs," *Journal of Research in Science Teaching* 24 (1987): 369-383.

Niemiec, Robert, and Walberg, Herbert J. "Comparative Effects of Computer-Assisted Instruction: A Synthesis of Reviews," *Journal of Educational Computing Research* 3, no. 1 (1987): 19-37.

Pelgrum, Willem J., and Plomp, Tjeerd. *The Use of Computers in Education World Wide—Results from the IEA Computers in Education Survey in 19 Educational Systems*. Oxford, England: Pergamon Press, 1991a.

Pelgrum, Willem J., and Plomp, Tjeerd. "The Use of Computers in Education Worldwide: Results from a Comparative Survey in 18 Countries." Paper presented at the Annual Meeting of the American Educational Research Association, Chicago, IL, 1991b.

Roblyer, M. D.; Castine, William H.; and King, F. J. "Assessing the Impact of Computer-based Instruction: A Review of Recent Research," *Computers in the Schools* 5, nos. 3/4 (1988): 11-149.

Roth, Kathleen J. *Conceptual Understanding and Higher Level Thinking in the Elementary Science Curriculum: Three Perspectives*. East Lansing, MI: Institute for Research on Teaching, Michigan State University, 1989.

Shymansky, James A., and Kyle, William C., Jr. "A Summary of Research in Science Education—1986," *Science Education* 72 (1988): 254-275.

Shymansky, James A.; Kyle, William C., Jr.; and Alport, Jennifer M. "The Effects of New Science Curricula on Student Performance," *Journal of Research in Science Teaching* 20 (1983): 387-404.

Striley Stein, Joanne. "The Computer as Lab Partner: Classroom Experience Gleaned from One Year of Microcomputer-Based Laboratory Use," *Journal of Educational Technology Systems* 15 (1986-1987): 225-236.

Suppes, Patrick, and Fortune, Ronald F. "Computer-Assisted Instruction: Possibilities and Problems," *NASSP Bulletin* 39, no. 480 (1985): 30-34.

Thijsen, J. Anke; van der Sijde, Peter C.; Collis, Betty; Plomp, Tjeerd; and Abbink, Marie-José. *De Effectiviteit Van Nascholing Informatietechnologie: Een Vooronderzoek* [The Effectiveness of Inservice Training in Information Technology: A Feasibility Study]. Enschede, The Netherlands: Department of Education, University of Twente, 1990.

van den Akker, Jan J. H.; Keursten, Paul; and Plomp, Tjeerd. "The Integration of Computer Use in Education," *International Journal of Educational Research* 17 (1992): 65-75.

van Schaick Zillesen, Piet G. "Methods and Techniques for the Design of Educational Computer Simulation Programs and Their Validation by Means of Empirical Research." Doctoral dissertation, University of Twente, Enschede, The Netherlands, 1990.

Voogt, Joke. "Courseware Evaluation by Teachers—An Implementation Perspective," *Journal of Computers in Education* 14 (1990): 299-307.

Walker, Decker F. "Computers and the Curriculum." In *Microcomputers and Education*, Eighty-fifth Yearbook of the National Society for the Study of Education, edited by Jack A. Culbertson and Luvern L. Cunningham, pp. 22-39. Chicago: University of Chicago Press, 1986.

Chapter 9

GENDER EQUITY

Lesley H. Parker, Léonie J. Rennie,
and Jan Harding

During the second half of the twentieth century, resources have
been provided in many parts of the world for development of curricula
in school science. While initially the rhetoric and practice of many of
these developments were directed to the need for more scientists, an
additional thrust toward the provision of "science for all" has developed
more recently. These directions reflect two of the overarching aims of
science education: (1) to educate students for careers in science and
technology; and (2) to create a scientifically and technologically literate
population, capable of looking critically at the development of science
and technology, and of contributing to democratic decisions about this
development.

In the context of these aims, concern has been expressed since the
early 1970s regarding the imbalance of female and male enrollments in
science classes. The concern has focused progressively on the consequences of this imbalance for universal scientific literacy, and for gender
equity in the pursuit of science and technology. It now is acknowledged
widely that by the end of schooling the pool of females with the motivation and background to progress into further studies in science, into
decision-making roles or careers in science and technology, and into
activities integral to the application of science and technology for development, is much smaller than the pool of males with similar motivation
and background. Clearly, if science education is to achieve the above
aims, it must provide more equitably than in the past for the access and

Chapter consultants: Jane Butler Kahle (Miami University, USA) and Shirley Sampson
(Monash University, Australia).

success of both females and males. The purpose of this chapter is to assist policymakers and practitioners in making such provisions.

We begin with research findings that focus on describing and attempting to explain the different patterns of participation by females and males in science. We then shift to educational policy and practice where we outline a number of strategies implemented in various parts of the world that address the problematic relationship between gender and science. The implicit assumption underpinning all of these strategies is that properly targeted, educationally based changes—involving, for example, curriculum materials, learning opportunities, classroom climate, and student-teacher interactions—can improve female students' participation and achievement in science and enhance their attitude toward science. From the strategies for change presented here, a number of features are drawn out that appear to be fundamental to the enhancement of gender equity in science, in terms of the two aims of science education stated at the outset of this chapter. In the final section of the chapter we present a summary and overview of the messages from research and evaluation, and emphasize the combinations of strategies that hold the most promise for effectiveness and efficiency in future policy and practice.

Wherever possible, we draw on work which itself constitutes a review or report of wide-ranging significance. In addition, smaller-scale research and vignettes from evaluation studies carried out by one of the authors (Harding, in press) are used selectively to provide particularly illuminating examples of features critical to the implementation and management of change in the gender and science relationship. In making these selections, we are aware of the many valuable documented efforts not cited and, in particular, of the many groups and activities for which limited data are available.

RESEARCH ON SCIENCE-RELATED SEX DIFFERENCES

In this section, we summarize research findings pertaining to a number of aspects of sex-related differences in science education. The focus initially is on research describing and analyzing the pattern of sex differences in enrollments and performance in the various science subjects and in different countries of the world. This kind of essentially descriptive data is fundamental to policy and practice at all levels of education from the classroom and school through to the national or international level. Indeed, the importance of having an adequate database relating to gender equity in science and technology has been remarked upon in many reports, particularly those focusing on developing countries. Thus, of the various messages from research to policymakers and practitioners in the chapter, the first is that *a lack of reliable data hampers the identification of both a problem and its solution.*

Given that any action in this area also needs to take account of basic research on various psychological attributes, the findings regarding sex differences in cognitive and affective abilities also are summarized in this section. We then shift attention to more sociologically based research that focuses on the image of science, especially the ways in which the image can be constructed in schools through the curriculum materials used and the kinds of learning environments created.

Enrollments in Science Studies

In 1981, Alison Kelly coined the term "the missing half" to describe the underparticipation of girls in science. Also in 1981, the first GASAT (Gender and Science and Technology) Conference was held in Eindhoven, in the Netherlands, with participation of representatives from eight different countries. All of these countries shared the problem of low enrollment of girls in science. Since that time, at the six subsequent GASAT Conferences and in other contexts, the worldwide scale and dimensions of this problem have been revealed. (See, for example, the reports of the two most recent conferences by Granstam and Frostfeldt, 1990, and by Rennie, Parker, and Hildebrand, 1991.)

In most countries for which data exist, including both developed and developing countries, the participation levels for females in science beyond the age at which the study of science is compulsory are lower than those for males. This sex difference tends to be very large for physics, somewhat small for chemistry, and either slight or nonexistent for biology. Thus the second message from research is that *once students are free to choose, females and males participate quite differently (in both quantitative and qualitative terms) in the study of science.*

Performance in Science

As in the case of science enrollments, there have been many studies, both large and small and in many different countries, of sex differences in science achievement. One of the most extensive was the 1970-1971 study of the IEA (International Association for the Evaluation of Educational Achievement). On the multiple-choice tests which formed the basis for this study, boys consistently performed better than girls in all nineteen countries sampled. The sex difference was shown to increase as students progressed through the school system, and to be greatest for physics, somewhat smaller for chemistry, and smallest in biology. The 1983 Second IEA Science Study (SISS) again involved administration of multiple-choice tests to samples of the 10-year-old, 14-year-old, and 17- and 18-year-old populations in participating countries. The sex difference revealed by the SISS was still in boys' favor, although smaller than that reported from the first IEA Study, and less consistent across countries (Keeves and Kotte, 1992).

Sex differences in science performance also were found by the Assessment of Performance Unit (APU) in the United Kingdom. Johnson (1987) reported that, in terms of the ability to apply physics concepts, the differences favoring males were small at age eleven and largest at age fifteen. In chemistry, smaller and inconsistent differences in favor of males were found, while in biology there was no consistent pattern. Differences in performance on practical skills were also heterogeneous, with males consistently performing better than females when handling certain instruments (such as ammeters) but not others, and females consistently performing better than males on tests involving observational skills. The APU reports point out that the patterns of sex differences revealed by tests of concept application and practical skills tend to correlate with students' reported past experiences. It appears that if a student can identify with the context of a test item or task, then a successful test outcome is more likely for that student.

Overall, the APU research reemphasizes findings reported elsewhere (AAUW, 1992) that highlight the importance of test and item context in relation to gender differences in performance, and it reinforces the idea that the kind of test used itself can be a variable affecting students' performance, with multiple-choice tests and decontextualized test items tending to favor boys. It raises again the issue of gender bias in testing, identified in quite a different way by Spear (1984), who showed that science work consistently received a higher mark from teachers if it was thought to have emanated from a male rather than a female, thus also indicating the extent to which teachers' grading practices appear to be contaminated by their gender-based expectations.

A further cautionary note regarding the interpretation of data on science performance is struck by the comprehensive statewide Western Australian data of Parker and Offer (1987). Focusing on sixteen successive cohorts of 15-year-old students (1972-1987), they show that, under a curriculum structure which provided identical science course taking for males and females, initial sex differences in males' favor shifted gradually to a point where there was a slight but consistent difference in favor of females. In this context, it is important to note that the first IEA study did not attempt to control for course-taking background of the students in the samples, although an attempt was made to control for "opportunity to learn" in the SISS. The absence of these controls possibly could account in part for the increase in science-related sex differences which the IEA studies identified in the higher years of schooling because in most school systems boys typically take more science (especially physical science) than girls, particularly in the later years of their schooling. Significantly, of all the countries included in the IEA studies, Thailand is the only one in which science is compulsory for virtually all students throughout secondary schooling, and it had the smallest science-related sex difference.

Thailand is also a country which has been the focus of some of the research on cultural influences on the achievement of males and females in science. Klainin and Fensham (1987) report that Thai girls perform at least as well as boys in physics and better than boys in chemistry. They propose that, in addition to the mandatory nature of the science curriculum in Thailand, cultural influences that associate chemistry with women's work, and the predominance of women among physical science teachers, have a positive influence on girls' achievement in the physical sciences. Other research from Nigeria, Jamaica, Poland, and Trinidad/Tobago cited in Rennie, Parker, and Hildebrand (1991) also suggests that boys and girls are more likely to perform equally in science when cultural norms and values reflect a society in which women traditionally have played an important economic role.

In summary, then, a third message from research is that *although earlier research appeared to establish a pattern of higher achievement by males than by females in physical sciences, particularly in the later years of schooling, the more recent research suggests that the situation is much more complex.* There is now less certainty attached to the concept of performance measurement, particularly in relation to the possibility of gender bias in the measures used. It is increasingly clear that trends in sex differences are affected strongly by cultural, societal, and educational changes, and that further investigation of all of these areas is required.

Cognitive Abilities Relevant to Performance in Science

Research on sex-associated differences in cognitive abilities has covered verbal, quantitative, and spatial abilities. Linn and Hyde (1989) provide a summary of much of this research, using meta-analysis to synthesize the findings of a large number of studies. Given that 1974 was the date of the influential Maccoby and Jacklin (1974) synthesis of sex differences, Linn and Hyde conducted their meta-analyses in two parts, examining research undertaken before 1974 and after 1974, to establish whether any changes had taken place in the fifteen-year period. Their findings are of particular interest. First, for verbal ability, they concluded that differences which previously were statistically significant are now negligible. Second, for quantitative ability, they demonstrated that sex differences have declined for most measures, and that those which remain are not consistent or are idiosyncratic to particular tests. Third, for spatial ability, which is cited frequently as an important factor contributing to sex differences in science achievement, their analysis indicated similarly that sex differences are declining and inconsistent and, furthermore, that those differences which remain are responsive to training.

Linn and Hyde's synthesis is one of the most significant recent contributions to this area of research. It provides a fourth message from research, which is that *contemporary data are required for decision making, and reliance on old, even established, findings is unsafe*. As in the case of research on performance in science, it is timely to remember the dynamic nature of the various phenomena under investigation and the importance of ensuring that educational strategies addressing gender equity in science are based on the most up-to-date and reliable research evidence in the area.

Affective Variables Relevant to Science

A comprehensive review of gender differences in affective variables was undertaken by Steinkamp and Maehr (1984). Although there was a small difference favoring males in overall attitude to science, when science was categorized according to biology, physics, or chemistry content, females were more positive than males toward biology and chemistry, but less positive toward physics. The findings of Steinkamp and Maehr suggest that sex differences in attitude toward science are small and, like achievement, vary with the subject matter and how the "attitude" is defined and measured.

A recurring factor in studies of attitude is the difference between males' and females' patterns of out-of-school science-related activities (Johnson, 1987; Sjøberg and Imsen, 1988). In general, females report more observational and biology-related activities, while males report more experiences with equipment and those kinds of activities which can benefit the later study of physical science. These differences match the pattern in enrollments in the branches of science when students begin to choose science subjects and make decisions about careers. Career-related decisions also have been linked to differences in self-image. Sjøberg and Imsen (1988) found that females described themselves as more empathetic and oriented toward others, and saw aspects of care and person-orientation in careers as relatively important. Males, however, described themselves as more competitive, perceived ego-orientation and personal benefit as important factors in career choice, and were far more likely than females to aspire to science/technical careers. All students perceived physics, of all the sciences, to be the least person-oriented career choice.

An important finding in relation to science attitudes comes from research showing that they can be changed by targeted activities. Parker and Rennie (1986), for example, showed that, while ten-year-old females were initially reluctant to be involved in constructing simple electric circuits, both their confidence and their attitudes in relation to this activity improved after involvement in it. This finding not only reinforces the significance of social and experiential influences on science outcomes, but it also indicates that planning courses around students' stated interests can preclude females from access to various experiences that they might enjoy.

In this context, a further important finding regarding attitude formation concerns the influence of parents. Eccles (1989) synthesized an extensive body of her group's research showing that parents hold strong gender-role stereotyped beliefs about their children's academic competencies. Irrespective of their children's demonstrated achievement, parents in the Eccles studies (1) expected sons to do better in mathematics than daughters (and with less effort), (2) saw mathematics, physics, and chemistry as more important for sons than for daughters, and (3) were more confident of sons' than of daughters' ability in mathematics and science. Importantly, parents' confidence is shown to have a direct impact on children's self-perception and values, partly through the experiences which parents provide for and encourage in sons and daughters.

The fifth message from research thus pertains to affective variables. It is that *affective gender differences are larger than cognitive differences, are more complex, and relate to students' attitudes toward science and their self-image with respect to science*. Two important subsets of this message are that *students' attitudes toward science can be improved by targeted activities and that students' self-perceptions in relation to science are influenced by parents' expectations*, which themselves probably are influenced by the public image of science.

The Image of Science

There is no doubt that science has a masculine image, and a considerable amount of the research on gender issues in science education has identified and attempted to explain this phenomenon (Kahle, 1988). In society at large, female scientists have had very low visibility. Young children thus tend to believe that scientists should be male. Girls are presented with little in the way of role models of successful females in science. Moreover, as Head (1985) has proposed following his synthesis of studies of the image of science and scientists, students' perceptions of scientists as male, emotionally reticent, unworldly, somewhat unsociable, and oriented to "things" rather than "people" could well discourage females from choosing to continue with science.

Indeed the importance of stereotyping of gender and of science has been demonstrated by Kelly (1988a) who examined the relationship between sex stereotypes and school science. This research was part of the longitudinal *Girls into Science and Technology (GIST)* project, implemented in the United Kingdom in 1980-1984. Kelly confirmed that females who hold strong sex stereotypes achieved less well in science, had worse attitudes toward science, and were less likely to choose branches of science associated with males. This tendency was stronger in females who were lower achievers and of low socioeconomic status. However, for males who overall exhibited stronger sex stereotyping than females, science attitude and option choice were not linked to sex stereotyping as strongly as for females, and strongly sex-typed males achieved less well in science than other males of similar general ability. This study provides an elegant and powerful illustration of the inhibiting effects of children's sex stereotyping on their academic development.

Thus the sixth major message from research is that *science has a strong masculine image which, in association with a general tendency toward sex-stereotyping, appears to be related negatively to females' choice of science and attitude toward science, especially in the cases of the females who are most at risk educationally, that is, those of lower ability and lower socioeconomic status.*

Schools and the Image of Science

Another synthesizing paper by Kelly (1985) explains how the masculine image of science is constructed in schools. First, school science is masculine in terms of the disproportionately large numbers of males who study and teach it. Second, the bias toward males in the way in which curriculum materials are presented and packaged offers an image of science as exclusive to males. Third, Kelly argues that the male-oriented pattern of classroom interaction contributes to the masculine image of science. The "numbers" dimension of Kelly's model now is supported by data from many sources (for example, the GASAT conference papers referred to earlier), and already has been highlighted in this chapter as part of the second message from research. The other "packaging" and "practice" dimensions of her model warrant further discussion.

The curriculum: Packaging and presentation. The instructional materials used in science classrooms now are known to carry implicit messages about the relationship between gender and science. Research reported by Whyte (1986) demonstrated that, particularly in the physical sciences, the illustrations, examples, and applications presented in resource materials are more familiar in general to the experiences and interests of males than to those of females. Science curriculum materials typically omit reference to women, and even can exclude the human and social aspects of science. Research has established the deleterious effects of such omissions on females' education in science and has led to recommendations for major reforms (AAUW, 1992).

From their extensive review of research into the gender characteristics of instructional materials, Scott and Schau (1985) concluded that exposure to sex-equitable materials encourages students to develop a gender-balanced knowledge of people, more accurate sex-role knowledge, and more flexible attitudes. In contrast, exposure to sex-biased materials both communicates and reinforces sex-biased expectations. Further, the use of male-generic language results in gender associations that are frequently male. The most gender-balanced associations result from the use of language which specifies both "he" and "she" rather than the gender-unspecified "they." Significantly, there is no evidence that sex-equitable materials decrease the comprehension or performance of males.

At a more philosophical level, others argue that scientific knowledge itself is masculine, and that this bias is exaggerated in the school curriculum. Harding (1986), for example, notes that the process of nurturance produces different emotional characteristics and cognitive styles in boys and girls, and that "science education has, in the past, been presented to match, rather exclusively, the needs of boys" (p. 5).

In summary, then, the seventh message from research is that *the way in which the curriculum presents science, and the images which are conveyed about who does science, who are scientists, and how science is practiced, all contribute to a picture of science which students construct as masculine.*

Learning environments: The practice of science teaching. The predominant pattern of teachers' interaction with their female and male students is now well established. As Kelly (1988b, p. 20) concludes from her meta-analysis of eighty-one studies of teacher-pupil interaction, it is now "beyond dispute that girls receive less of the teacher's attention in class, and that this is true across a wide range of different conditions." Kelly identifies science as one of the areas in which female students are particularly underinvolved in lessons. It is important to note that the conditions to which Kelly refers encompass all age groups, different countries, various socioeconomic and ethnic groups, all subjects in the curriculum, both male and female teachers, both pupil-initiated and teacher-initiated interactions, and all major categories of classroom interaction (for example, behavioral criticism, instructional contacts, high-level questions, academic criticism and praise). It is also important to note her finding that "teachers trained in sex equity distribute their attention equally between the sexes in everything except criticism" (Kelly, 1988b, p. 21).

There are gaps to be filled in this research, however. Again, as pointed out by Kelly, despite the wealth of descriptive research, there is as yet no research that establishes a clear link between patterns of class-room interaction and sex differences in achievement and attitudes toward science. There is a need for future research to focus on whether or not a causal relationship between these variables exists. Work carried out in mathematics education provides some relevant findings in this regard. Stage et al. (1985), in a synthesis of the work of Eccles and others, found at most only a very weak overall relationship between student-teacher interaction patterns and mathematics-related attitudes and future study plans. They also report, however, that studies of the impact of a single salient teacher suggest that a teacher's impact on female students' attitudes toward mathematics is large if the teacher provides them with active encouragement. Such encouragement is seen to come from exposing the females to role models, praising them sincerely for high performance, and giving them explicit advice regarding the value, especially career-related value, of mathematics. In support of this finding, Kahle (1988) reports that biology teachers who were successful in encouraging both females and males to continue studies in science also offered, among other things, encouragement and career-related advice.

In summary, this powerful and complex eighth message is that *there is an unconscious bias of teacher attention toward males in a wide variety of educational settings, but especially in science classrooms, and it hints at the positive effect of individual, gender-aware science teachers upon females' attitudes toward science and the likelihood of their continuing in science.*

Learning environments: Student-student interaction. In recent years, there has been increased recognition that much of what transpires in schools and classrooms is of a competitive nature, and that not all students, particularly not all female students, thrive in such an environment. From their extensive study of Australian students' preferred learning environments, Owens and Barnes (1982) report that, at each grade level from Grade 4 through to Grade 12, females expressed a stronger preference for cooperative learning, while males expressed a stronger preference for competitive learning and for individualized learning. They suggest that these preferences reflect early socialization of males and females, with that of females emphasizing cooperative relationships and mutual assistance, and that of males emphasizing adventurous striving and competition. They also suggest that science classrooms present a particularly competitive image which might not be conducive to females' participation and achievement in science. A considerable body of research now has established that cooperative learning strategies have a positive effect on students' achievement and attitude, especially in science (Eccles, 1989).

The ninth message thus *is that the difference in males' and females' preferred learning styles and environments points to a need to restructure the currently highly competitive learning environments typical of most schools to include opportunities for students to work and be assessed in a cooperative, gender-inclusive group context.*

Learning environments: Mixed or single-sex classes? The relative advantages and disadvantages of mixed-sex and single-sex learning environments have been debated for many years. The debate has been popularized to a considerable extent, and many of the current assertions about the two environments do not have a sound basis in rigorous scholarly research. Although there is some research focusing on schooling in general, few studies have addressed the effects of single-sex or mixed-sex organization (whether within a school or classroom, or in a school as a whole) specifically in relation to science outcomes.

Generally, the proponents of single-sex grouping hold that sex segregation is a solution to the gender-based inequities found in coeducational classrooms and schools, citing evidence that single-sex groupings appear to reduce the degree to which females see physics and chemistry as masculine, and to bring about a small improvement in females' physics achievement (see Kelly, 1985). Other large-scale studies, however, provide somewhat contradictory evidence and suggest that other factors (such as mode of assessment, school organization, and teaching styles) also were influencing outcomes.

The tenth message from research, then, is that *overall the research evidence in relation to the single-sex/coeducation debate is mixed. To date, however, research does not justify attempts to improve science outcomes for females simply by teaching them in single-sex groups.*

RESEARCH ON STRATEGIES FOR ELIMINATING SCIENCE-RELATED SEX DIFFERENCES

International Perspectives on Interventions

During the past fifteen years or so, projects aimed at enhancing gender equity in science have been implemented in many different countries of the world and at all levels of education, from early childhood education through to graduate education of teachers (including initial training at the graduate level), and in informal educational settings such as museums. Perusal of papers presented at recent GASAT conferences (Granstam and Frostfeldt, 1990; Rennie, Parker, and Hildebrand, 1991) and chapters in Parker, Rennie, and Fraser (in press) reveal the following worthwhile activities in this area:

- large nationwide government-funded projects such as the JETS (Junior Engineers and Scientists) in Nigeria;
- small government-funded interventions, like those provided for educators by the Ontario Women's Directorate in Canada;
- national policy initiatives, such as in Australia and New Zealand, that incorporate gender equity into national statements on science education, and the development and dissemination of national policies on the education of females in science and mathematics;
- national curriculum initiatives, like Denmark's revision of physics and chemistry courses to take account of the preferred learning styles of females;
- university-based projects such as the Girls, Physics and Technology (MENT) project at the Eindhoven University of Technology in the Netherlands, and the Norwegian Primary Science Project at the University of Oslo;
- locally based (and in part privately sponsored) initiatives such as the training program for teachers conducted by the science and technology center (Teknikens Hus) in Luleå in Sweden, and the project using role models in Bhopal City, India;
- privately funded projects such as the Van Leer Jerusalem Institute *Naaleh* Program, which is a comprehensive, multiphased intervention aimed at increasing the number of females prepared at school level to enter science and engineering studies.

Because of all of this activity, the information base relating to intervention projects has expanded rapidly during the past decade. Although, for a variety of reasons, many of the initiatives have not been evaluated, a number of critical features emerge for which evidence is available from comprehensive and systematic evaluation. To illustrate these points, the next section provides an overview of evaluations conducted in the United States.

Evaluation of Interventions: Emergence of Critical Features

Stage et al. (1985) estimate that during the 1970s and early 1980s as many as 600 programs aimed at improving the quality and quantity of science and mathematics education for females were developed and implemented in the United States. The programs were funded from many different sources. Some focused on careers, with the assumption that mathematics and science were necessary "filters" for entering many careers. In general, the programs were voluntary and thus likely to attract young people with a particular commitment to upgrading their mathematics and science.

For some of the 600 programs descriptive and evaluative data are available, together with judgments regarding possible replicability and wider impact. The discussion of these programs by Stage et al. (1985) focuses on six categories embracing: (a) special classes for females (for example, "Math for Girls" provided by the Lawrence Hall of Science at the University of California at Berkeley and the "Women in Engineering" initiative of Purdue University); (b) special classes aimed at addressing problems faced by females (for example, mathematics anxiety); (c) curricula designed to address special needs of females (for example, COMETS—Career Oriented Modules for Exploring Topics in Science); (d) teacher education programs; (e) school district-based efforts (including a variety of available materials, expertise, and role models); and (f) extracurricular or cocurricular activities (for example, visiting programs, conferences, and support networks).

The overall conclusion of the Stage et al. evaluation of these programs is that success is associated with three features: a strong academic emphasis; multiple strategies; and a systems approach. Importantly, this evaluation suggests that these elements, while representing the strengths of programs for increasing females' participation and achievement in mathematics, science, and engineering, also represent sound educational practice in relation to all students.

Malcom (1984) makes similar points in her report of the evaluation of 167 programs in the United States aimed at facilitating increased access and achievement among females and/or minorities in K-12 mathematics and science education. The evaluation team identified "exemplary" projects and described their characteristics. Criteria used to assess exemplariness included a program's achievement of stated goals, its duration, its ability to attract outside support, its popularity (that is, the ratio of applicants to participants), its reputation with local scientists from affected groups, its cost effectiveness, and how readily it had been copied. Malcom's report provides an invaluable basis for future action in this area. Its seven highly practical recommendations highlight a number of critical elements, which are supported and illustrated by findings from elsewhere and are discussed in the next section.

Generalizability of Critical Elements

The need for specific targeting. One of Malcom's findings was that "unless programs 'for all' specifically assess the status of, articulate goals for, and directly target educational problems of females . . . they are unlikely to be effective" (p. xiii). This is well-illustrated by an example from a coeducational school in the United Kingdom. In this school, the Head of Science offered a special "girl-friendly" science course and careers program to female students in the third year of the secondary school (the year in which, traditionally, pupils made their choices for public examination courses). The special careers program involved the students in meeting women scientists and technologists, visiting engineering works, and participating in counseling discussions. The combined curriculum/counseling approach met with some success in that the numbers of females choosing examination courses in physics and chemistry showed a marked increase. The teacher then was persuaded to introduce a broad-based science course into the school at the third-year level. It was assumed that, if the materials of this course were "girl-friendly," no special counseling program in relation to later choices would be necessary. However, when the first group of students was required to choose advanced-level courses, fewer females than in previous years chose science subjects, thus pointing to the need to reinstate the special targeting of females.

It is apparent from this example that merely making a subject compulsory "for all" does not necessarily result in females feeling included, even when the curriculum materials are tailored to their perceived needs. The females in this case needed to hear specific messages that women could and did work successfully in what appeared to them to be traditionally male areas, and that opportunities were there for them to do the same.

Indeed, the advantages of targeting are demonstrated by an example from Karaikudi in India (Granstam and Frostfeldt, 1990). Here the Center for Women's Studies at Alagappa University "adopted" a village and targeted female dropouts who, for financial reasons, were unable to continue attending secondary school. Some government funding enabled a stipend to be paid to them while training at the university in a course focusing on the repair of domestic electrical appliances. Although initially reticent about becoming involved in an area perceived by them to be more appropriate for males, the trainees quickly adapted to this practical course, completing it with marketable skills in repairing appliances and, importantly, with some status in their community.

The risks of mainstreaming. A second of Malcom's recommendations related to mainstreaming, which she considered to be both possible and desirable, "but only after specific targeting, followed by institutionalization of program elements critical to the achievement of . . . females and monitoring to assure that participation levels are maintained" (p. viii). Indeed, the pitfalls of premature and unsupported mainstreaming have been demonstrated in a number of contexts. The British Technical and Vocational Education Initiative (TVEI), funded and managed centrally by the Manpower Services Commission, is a particularly interesting example. It focused on 14- to 18-year-old students across the whole ability range, and had explicit links to gender equity and vocational education. A study of the implementation of TVEI in Hertfordshire (James and Young, 1989), however, revealed that long-term benefits appeared to be in jeopardy unless the implementation was planned to sustain the life of the initiative beyond the initial project funding (four years in this case).

A further example comes from Western Australia. As indicated earlier in this chapter, Parker and Offer (1987) have documented the equal science achievement of 15-year-old females and males, under the systemwide curriculum structure that operated between 1969 and 1987 and required all students to study the same multidisciplinary science curriculum for the first three years of secondary schooling. Subsequently, however, Rennie and Parker (1993) demonstrated that these gains were dissipated under a new structure introduced in 1988, which allowed students considerably more choice regarding both the amount and the kind of science that they studied at the lower secondary level. Under the new, less-prescribed structure, there was a reduction, particularly among females, in the total amount of science studied and in the emergence of strong sex stereotyping in students' choice of science topics. Again, in a main-streamed initiative, even one such as this with the explicit goals of excellence, equity, and relevance, "equity" was not translated from the rhetoric into the reality without specific measures to provide support for females.

The need for systemic support. Malcom's evaluation also demonstrated the fundamental importance of systemic support for equity initiatives. Again, this is supported by examples from elsewhere in the world. Harding (in press) reports the case of a college of higher education in the United Kingdom, where a female senior lecturer, with help of colleagues in engineering, set up a day of hands-on technology activities for female students aged thirteen and fourteen years. Each secondary school in the county was invited to send a given number of female students and arrangements were made to transport them to the college. The exercise, scheduled during time allocated to examination marking, required considerable organizational effort on behalf of staff and especially of the coordinator. Senior administrators gave verbal support to the initiative, but the time which staff gave was not credited to their teaching loads and no additional remuneration was awarded. Despite the demonstrable success of this initiative, systemic support never was forthcoming and, without this, the project ultimately was abandoned.

This example can be contrasted with an example taken from the state of Victoria in Australia, where a grass roots development was supported by system-based policy and resources. The Victorian initiative began in 1983 when a network of science teachers established the McClintock Collective, following an in-service course which addressed equal opportunities in the science classroom. The Collective articulated a commitment to promoting science education that is relevant to and inclusive of females of all ages and backgrounds. The original membership of twelve expanded rapidly to include more than 400 active or interested members. Over the years, government support for the network has been forthcoming in a number of ways. On a regular basis, the Collective plans and runs state-supported professional development courses and workshops for other teachers. It has been provided with funding to train its own members to enable them to develop greater confidence and skills in running workshops, in facilitating the equal participation of others, and in handling conflict in controversial areas. Funds also have been made available to cover the employment of a project officer.

The work of the Collective has flourished in the context of clear policy statements on females' education by both the federal and state governments. Indeed, members have contributed to the development of these statements. Further, a collection of McClintock "gender-inclusive" strategies has been published by the national curriculum organization, and a national project to create, try, and publish gender-focused professional development materials for science and mathematics teachers was awarded to the Collective (Lewis and Davies, 1988). In mid-1990, a Mathematics, Science, Technology Education Center for Girls was established in Victoria to include the McClintock Project Officer and three other workers. The developments surrounding the McClintock Collective thus demonstrate a unique and productive coming together of official policy and resources and of grass roots commitment.

The need for strong leadership and committed teachers. Malcom's evaluations also identified leadership and commitment of personnel as critical to the success of initiatives. Again, reports of other projects highlight these critical elements. For example, the evaluation of TVEI referred to earlier in this chapter suggested that, in order to incorporate gender equity systematically in schools operations, well-targeted teacher in-service activities and the development of networks of concerned and committed school staff were required (James and Young, 1989). A similar point was made in the reports of the three-year GIST project (Whyte, 1986) which was referred to earlier.

The GIST project's approach to working with teachers focused on their professional commitment to providing good learning experiences for all their pupils. The project team worked with teachers to develop and implement nonsexist curriculum materials and teaching strategies. At the end of the project's three years, it was found that, although the pattern across all targeted schools showed no statistically significant change, changes in the pattern of girls' subject choices occurred in individual schools. Importantly, the changes in subject choice patterns and in attitudes varied according to the level of commitment of schools to the aims of the project and the extent to which administrators and teachers were prepared to take a leadership role in this regard. The changes were generally greater in schools that had embraced the project wholeheartedly, using its resources to support and extend a commitment already articulated, than in schools that had reacted with polite tolerance, resentment, or hostility.

Thus, one of the most significant insights to emerge from GIST is the importance of linking policy, through personal and material resources, to grass roots commitment in order to achieve significant changes in expectations and behavior. It highlights the importance of commitment from both teachers and management in schools and of the need for continuing in-service teacher education.

The need for sustained, long-term programs. While there are many instances of the success of short-term initiatives for motivational, awareness-raising, or morale-boosting purposes, there is strong evidence that, in terms of both concept and resources, initiatives need to be sustained. The comprehensive report of Åsekog (1986) of activities implemented in Sweden to encourage girls into the science/technology stream of upper secondary education provides a compelling illustration of this point. In association with a special information program directed at ninth grade girls in Stockholm, the number of girls entering the science/technology stream increased steadily by nearly 300 percent from 1974 to 1980. By 1980, because there was a feeling that the program was well established, its resources were reduced, and this was accompanied by a 25 percent decrease in the number of girls selecting the target stream. With renewed resources and effort in 1982, however, the program's earlier rate of success was reestablished. This example illustrates what is perhaps one of the most important points highlighted by Malcom's evaluation, namely, the need for intervention programs to be sustained and long-term.

SUMMARY, IMPLICATIONS, AND RECOMMENDATIONS

We began this chapter with the premise that all young people should be given the opportunity to be part of the pool of future scientists and technologists and to be scientifically literate citizens. Because females and males appear to possess equal potential to develop skills required for the pursuit of science, it is both a waste of talent and a deprivation to individuals that the two sexes do not participate equally in science.

Decision making which might address the factors responsible for the sex imbalance in science participation should take account of some of the following messages which emerged from the research reviewed in this chapter.

- When free to choose, males and females make different choices with respect to science.
- Sex differences in performance are decreasing.
- There is a need for contemporary and reliable data.
- Gender differences on affective variables are more pervasive than those pertaining to cognitive differences and they are amenable to change.
- Science has a masculine image which is detrimental to females.
- The masculine image of science is conveyed in the way in which the curriculum is packaged and science is taught and assessed.
- Teachers can perpetrate the myth that science is for males through patterns of classroom interactions and by creating competitive learning environments.

Attempts to remedy the gender imbalance in participation in science must focus primarily on affective, cultural, and social aspects, particularly as they are reflected in schools and curricula. Our discussion of reviews and evaluations of strategies and interventions suggests several clear, recurring, and compelling messages:

- Intervention programs need to be targeted at specific problems in order that the audience can be identified properly and appropriate goals articulated.
- Support mechanisms are needed for the staff involved.
- Consistent support from the educational system of which the initiative is part is necessary for both the implementation and continuance of an intervention program.
- Strong leadership and committed teachers are needed.
- Intervention programs need to be sustained for a longer term.

The overall message from these findings for policymakers and practitioners wishing to enhance gender equity in science education concerns both policy and support. A policy statement addressing the issues of equity in science education should be formulated at the highest level of responsibility for education in a community, state, or country. However, no policy can become effective without appropriate structures and support directed to its implementation. Effective implementation of equity policy requires the recurrent allocation of resources for action concerning the curriculum, the provision of material and human resources, the monitoring of change, and collaboration with schools and community groups.

The school curriculum needs to be structured to facilitate the achievement of the objectives of the policy. Because gender biases influence the choices made, science needs to appear as a compulsory subject in the school curriculum. Further, because males and females choose different science topics, the science course prescribed should be broad-based, including physics, chemistry, biology, astronomy, geology, and health. Further, it is desirable also that the curriculum structure allows time for young people to reflect on, and to challenge, gender stereotypes within their culture. To this end, a sensitive personal and social curriculum would be an ongoing strand within compulsory education, and would include substantial career education directed at breaking down gender-stereotyped expectations.

Curriculum development must be informed by the research evidence because gender bias can enter curriculum materials through the type of language used, the choice of examples, the background experiences, interests, and motivations that are assumed, the learning style implied, and the way in which the subject field is projected. In particular, it would be desirable for career education materials to include case studies of successful women scientists and technologists and of young women recruits who are enjoying the early stages of such work. Further, to facilitate policy implementation, education programs could be established to enable administrators, school governors, teacher educators, and teachers, first, to become aware of the ways in which gender stereotyping can cause disadvantage in science education and, second, to develop skills to help them counter gender stereotyping. For example, the study of gender and education interactions could form a component of teacher education courses at both preservice and in-service levels.

Systematic monitoring of student participation and outcomes is an important aspect of policy implementation. Data on any sex differences in achievement in science and course participation at all levels of academic, professional, and vocational education should be collected systematically and published. Further, because assessment of achievement in science has been shown to have strong gender interactions in the monitoring of participation and achievement, the links between outcome measures, assessment types, and gender must be recognized and taken into account. The results of monitoring could provide useful information for curriculum development and revision and for the design of teacher-constructed tests and formal external examinations. Where the education system employs inspectors or advisers to monitor educational practice, such persons need to be aware of gender interactions with teaching, learning, and assessment.

Based on the above discussion, the following recommendations for improving gender equity in science education are made:

1. A broad-based science course, including physics, chemistry, biology, astronomy, geology, and health, should be compulsory in the school curriculum.
2. The school curriculum should include consideration of gender stereotypes and career education aimed at breaking down these stereotypes.
3. Resources in science education should be gender inclusive, avoid gender bias in language and choice of examples, and include case studies of successful women scientists.
4. Science teachers, administrators, teacher educators, and teacher trainees should undertake educational programs that make them aware of the problems of gender stereotyping and give them the skills to counter it.
5. In order to guide attempts to overcome gender inequity, data on sex differences in student participation and outcomes should be collected systematically and published.
6. Gender bias should be avoided in the content, context, and mode of assessment in science.

We wish to emphasize two points in summing up this chapter, the aim of which is to assist policymakers and practitioners to provide equity in access and success for females and males in science. First, at all levels of education, policy is an essential prerequisite to systematic action. Second, policy alone is not enough. It must be supported by clear commitment from senior management levels and by the allocation of resources on a recurrent basis. Policy and the provision of resources must operate jointly to ensure that school curricula and supporting materials are informed by the now considerable knowledge about how girls and boys learn science, and how they choose to continue with science. Fundamental in this regard is compulsory study of a gender-inclusive, broadly based, multidisciplinary science curriculum, together with substantial, nonsexist career education and structured opportunities for personal and social development, particularly in relation to the interaction between science and society. In addition, policy and resources must ensure that student participation and learning in science are monitored systematically, using techniques that are fair and equitable to all students. Above all, and again at all levels, they must provide effective professional development about gender issues for everyone engaged in the education enterprise.

The researchers and practitioners whose efforts have been recorded in this chapter were concerned primarily with better education in science for all students. Thus, as indicated earlier (Stage et al., 1985; Whyte, 1986), it is being recognized increasingly that the strategies recommended here in the interests of gender equity in science are sound educational strategies for all students, whatever their sex and background.

REFERENCES

Åsekog, Brit. *Activities to Promote Equality of Opportunity for Girls in Technical and Vocational Education.* Paris, France: UNESCO, 1986.

AAUW (American Association of University Women). *How Schools Shortchange Girls.* Commissioned report prepared by the Wellesley College Center for Research on Women. Washington, D.C.: AAUW Educational Foundation/National Education Association, 1992.

Eccles, Jacquelynne. "Bringing Young Women into Math and Science." In *Gender and Thought: Psychological Perspectives*, edited by Mary Crawford and Margaret Gentry, pp. 36-58. New York: Springer-Verlag, 1989.

Granstam, Ingrid, and Frostfeldt, Inger, eds. *Contributions and Reports Book: European and Third World GASAT Conference.* Jönköping, Sweden: Jönköping University College, 1990.

Harding, Jan, ed. *Perspectives on Gender and Science.* London, England: Falmer Press, 1986.

Harding, Jan, ed. *Strategies in Action*. Report to the Equal Opportunities Commission. London, England: Equal Opportunities Commission, in press.

Head, John. *The Personal Response to Science*. Cambridge, England: Cambridge University Press, 1985.

James, Carol, and Young, Jane. "Case Study: Equal Opportunities Through the Hertfordshire TVEI Project." In *Changing Perspectives on Gender*, edited by Helen Burchell and Val Millman, pp. 14-28. Milton Keynes, England: Open University Press, 1989.

Johnson, Sandra. "Gender Differences in Science: Parallels in Interest, Experience and Performance," *International Journal of Science Education* 9 (1987): 467-481.

Kahle, Jane B. "Gender and Science Education II." In *Development and Dilemmas in Science Education*, edited by Peter J. Fensham, pp. 249-265. London, England: Falmer Press, 1988.

Keeves, John P., and Kotte, Dieter. "Disparities between the Sexes in Science Education: 1970-84." In *The IEA Study of Science III*, edited by John P. Keeves, pp. 141-164. Oxford, England: Pergamon Press, 1992.

Kelly, Alison. "The Construction of Masculine Science," *British Journal of Sociology and Education* 6 (1985): 133-154.

Kelly, Alison. "Sex Stereotypes and School Science: A Three Year Follow-Up," *Education Studies* 14 (1988a): 151-163.

Kelly, Alison. "Gender Differences in Teacher-Pupil Interactions: A Meta-Analytic Review," *Research in Education* 39 (1988b): 1-23.

Klainin, Sunee, and Fensham, Peter J. "Learning Achievement in Upper Secondary School Chemistry in Thailand: Some Remarkable Reversals," *European Journal of Science Education* 9 (1987): 217-227.

Lewis, Sue, and Davies, Anne. *Gender Equity in Mathematics and Science*. Canberra, ACT: Curriculum Development Center, 1988.

Linn, Marcia C., and Hyde, Janet S. "Gender, Mathematics and Science," *Educational Researcher* 18, no. 8 (1989): 17-19 and 22-27.

Maccoby, Eleanor, and Jacklin, Carol. *The Psychology of Sex Differences*. Stanford, Calif.: Stanford University Press, 1974.

Malcom, Shirley M. *Equity and Excellence: Compatible Goals*. Washington, D.C.: American Association for the Advancement of Science, 1984.

Owens, Lee, and Barnes, Jennifer. "The Relationship Between Cooperative, Competitive and Individualized Learning Preferences and Students' Perceptions of Classroom Learning Atmosphere," *American Educational Research Journal* 19 (1982): 182-200.

Parker, Lesley H., and Offer, Jennifer A. "School Science Achievement: Conditions for Equality," *International Journal of Science Education* 9 (1987): 263-269.

Parker, Lesley H., and Rennie, Léonie J. "Sex-Stereotyped Attitudes about Science: Can They Be Changed?" *European Journal of Science Education* 8 (1986): 173-183.

Parker, Lesley H.; Rennie, Léonie J.; and Fraser, Barry J., eds. *Gender, Science and Mathematics: Shortening the Shadow*. Dordrecht, The Netherlands: Kluwer, in press.

Rennie, Léonie J., and Parker, Lesley H. "Curriculum Reform and Choice of Science: Consequences for Balanced and Equitable Participation and Achievement," *Journal of Research in Science Teaching* 30 (1993): 1017-1028.

Rennie, Léonie J.; Parker, Lesley H.; and Hildebrand, Gail M., eds. *Action for Equity: The Second Decade*. Perth, Western Australia: National Key Center for School Science and Mathematics, Curtin University of Technology, 1991.

Scott, Kathryn P., and Schau, Candace. "Sex Equity and Sex Bias in Instructional Materials." In *Handbook for Achieving Sex Equity Through Education*, edited by Susan S. Klein, pp. 218-232. Baltimore, Md.: Johns Hopkins University Press, 1985.

Sjøberg, Svein, and Imsen, Gunn. "Gender and Science Education I." In *Development and Dilemmas in Science Education*, edited by Peter J. Fensham, pp. 218-248. London, England: Falmer Press, 1988.

Spear, Margaret G. "Sex Bias in Science Teachers' Ratings of Work and Pupil Characteristics," *European Journal of Science Education* 6 (1984): 369-377.

Stage, Elizabeth K.; Kreinberg, Nancy; Eccles (Parsons), Jacquelynne; and Becker, Joanne R., with others. "Increasing the Participation and Achievement of Girls and Women in Mathematics, Science and Engineering." In *Handbook for Achieving Sex Equity through Education*, edited by Susan S. Klein, pp. 237-268. Baltimore, Md.: Johns Hopkins University Press, 1985.

Steinkamp, Marjorie W., and Maehr, Martin L. "Gender Differences in Motivational Orientations Toward Achievement in School Science: A Quantitative Synthesis," *American Educational Research Journal* 21 (1984): 39-59.

Whyte, Judith. *Girls into Science and Technology: The Story of a Project*. London, England: Routledge & Kegan Paul, 1986.

Chapter 10

CROSS-NATIONAL
COMPARISONS OF OUTCOMES
IN SCIENCE EDUCATION

John P. Keeves

Each country needs an ample supply of high school graduates with an adequate knowledge of science to sustain its scientific and technological development and its economic well-being. It is also important that the members of the general workforce possess the knowledge and skills that permit them to cope with the increasing scientific and technological complexity of the devices employed in industry, the office, and daily life. Moreover, it is essential that, in a modern democracy, the enfranchised adult population understands the issues associated with the many complicated social problems that have arisen from recent scientific and technological advances. This chapter, however, is concerned primarily with the questions which arise in ensuring that a country produces a sufficient number of people who are trained professionally in scientific and technologically related fields to maintain its productive capacity, its economic growth, and a rising standard of living.

In this chapter I am concerned with the outcomes of science education rather than the inputs or the processes which operate. In particular, I am concerned largely with the outcomes achieved at the end of twelve years of schooling in terms of performance on science tests, the holding of favorable attitudes toward science, and the desire and expectation to continue with further study of science and to train for entry into a science-related professional occupation. However, there are intermediate

Chapter consultants: T. Neville Postlethwaite (University of Hamburg, Germany) and K. C. Cheung (University of Macau).

goals in science education that must be achieved satisfactorily by a school system to ensure a high level of outcome at the terminal secondary school stage. Hence, consideration is given also to the factors that influence the outcomes of science education at the upper elementary or 10-year-old level, and at the lower and middle secondary school or 14-year-old level. Moreover, attention is given to how the outcomes at these intermediate stages relate to outcomes at the terminal stage.

The primary data presented here are derived from the Second IEA Science Study (SISS), which was carried out by the International Association for the Evaluation of Educational Achievement in twenty-three countries during the years 1983-1984. The main topics covered in this chapter are: participation and achievement rates in science in different countries; associations between science achievement, time spent on learning science, and opportunity to learn the content in the science test; and factors such as home background, sex, and aptitude that affect science participation and achievement.

For a more detailed account of these studies, the international reports by Rosier and Keeves (1991), Postlethwaite and Wiley (1991), and Keeves (1992) can be consulted. Furthermore, each of the participating countries has produced national reports that address issues of particular concern within that country. These are the only comparative data available that seek to identify and explain the influence of a wide range of factors on the outcomes of science teaching and learning for a large number of countries in different parts of the world.

PARTICIPATION IN SCIENCE AT THE TERMINAL SECONDARY SCHOOL LEVEL

In the main, those high school graduates who undertake the further study of science after leaving school and train for entry into science-related professional occupations have been involved in the study of science at the final stage of secondary schooling. An important prerequisite for the production of an ample supply of high school graduates with an orientation toward science is that there should be at least an adequate proportion of the age cohort enrolled at school at the terminal stage of secondary schooling. It is also essential that large enough proportions of these students are enrolled in science courses in the foundational fields of biology, chemistry, and physics.

Table 10-1 records, for those countries that took part in the Second IEA Science Study (SISS) at the terminal secondary school level, the proportions of the age cohort enrolled at school and participating in specialist science courses. In some countries, educational provision at the terminal secondary level is subdivided into academic and vocational schooling. In the SISS, only those students in academic and general schools were tested, because students in vocational schools were not taking courses in science that would lead on to further study at a university or institute of higher education. In 1983-1984, countries differed markedly in the proportion of the age cohort enrolled at school at the terminal secondary level; the range was from 89 percent in Japan, 83 percent in South Korea, and over 80 percent in the USA, to 20 percent in England, 17 percent in Singapore, and 1 percent in Ghana. It should be noted that, in Japan and South Korea, substantial proportions of the age group (26 percent and 45 percent respectively) were enrolled in vocational schools.

The proportions of the age cohort in each country enrolled in specialist science courses in the fields of biology, chemistry, and physics differed markedly, with sizeable proportions in such programs in Australia, Canada, Finland, Hong Kong, Israel, Japan, and South Korea (table 10-1). This raises questions regarding the appropriate proportion of students required by a country to be enrolled in science courses in order to obtain an adequate number of high school graduates who might train for entry into science-related professions. Countries such as Australia, England, Sweden, and the USA currently are known to be experiencing an undersupply of professional workers in the fields of science and technology. In part, this shortfall could be due to the comparatively low level of financial reward offered to such professional workers. No doubt it is due also to a failure to ensure that ample numbers have graduated from university courses in scientific and technological fields over recent decades. It should be noted that Japan, South Korea, and Hong Kong appear to be well placed to satisfy a growing demand for scientifically and technologically trained personnel, provided that training is available for sufficient numbers in higher educational programs.

ACHIEVEMENT IN SCIENCE

Previous studies conducted by IEA in the areas of mathematics (Husén, 1967) and science (Comber and Keeves, 1973) have established

Table 10-1

Retention and Participation Rates and Mean Achievement Test Scores for Terminal Secondary School Students, 1983-1984

Country	% Retention at Terminal Secondary Level			% Participation and Mean Achievement for Fields of Science						
	Total %	Academic %	Vocational %	Biology Specialist %	Mean[c]	Chemistry Specialist %	Mean[c]	Physics Specialist %	Mean[c]	Non-science Specialist %	Mean[c]
Australia	39	39		18	53.4	12	53.5	11	53.6	10	60.1
Canada (English)	68	68		28	48.7	25	43.1	18	47.0	na	56.8
Canada (French)	79	67	12	7	42.3	37	32.1	35	30.2	na	51.0
England	20	20		4	38.1	5	73.1	6	64.2	10	65.2
Finland	59	41	18	-	-	16	40.5	14	44.7		
Ghana	1	1		0.2	63.6	0.6	63.7	0.6	54.3		
Hong Kong (Form 6)	27	27		12	61.0	20	66.9	20	64.4		
Hong Kong (Form 7)	20	20		7	70.2	12	78.5	12	73.7		
Hungary	40	18	22	3	65.6	1	53.4	4	62.7	10	67.5
Israel	65	65		20	56.9	8	50.3	12	54.4	na	49.1
Italy	34	34		4	46.7	1	42.9	13	34.2	21	58.9
Japan	89	63	26	12	51.3	16	57.8	11	61.8	35	66.2
Korea	83	38	45	38	44.5	37	33.9	14	43.3		
Norway	40	40		4	59.9	6	47.9	10	56.5	na	64.7
Poland	28	28		9	59.1	9	49.5	9	56.4		
Singapore	17	17		3	70.9	5	68.7	7	59.2	na	56.2
Sweden	28	28		5	60.5	13(6)[d]	47.3	13	51.4	15	67.1
Thailand	29	14	15	7	47.2	7	38.8	7	35.9	7	50.3
USA(A) (1983-84 testing)	80[e]	80[e]						12	35.0[e]	49	
USA(B) (1986 testing)[f]	83[e]	83[e]		12	41.8[e]	2	40.1[e]	1	49.2[e]		

Estimated regression parameters	a	b	r	n	Outliers (well above or below expectation) excluded from calculations
Biology	68.8	-0.62	-0.85	14	Canada (Fr.), Italy, Thailand, USA(B)
Chemistry	61.8	-0.79	-0.78	14	Hong Kong (F6 and F7), Italy, Thailand, USA(B)
Physics	66.9	-1.13	-0.91	13	Ghana, Hong Kong (F6 and F7), Italy, Thailand, USA(A) and USA(B)

[e]	Estimated value	[a]	Intercept
[d]	Sweden tested 13% of age group; only 6% were studying chemistry.	[b]	Slope of regression line
[c]	Mean percentage correct on 30-item test, not corrected for guessing	[r]	Product moment correlation
na	Data not available	[n]	Number of countries used in calculation
[f]	The USA conducted two testing programs. The 1983-84 testing is referred to as USA(A) and the 1986 testing as USA(B).		

that, at the terminal secondary school stage, the mean level of achievement in mathematics and science across countries is related inversely to the proportion of the age group enrolled in the study of the subject. Consequently, with substantial differences between countries in the proportions of the age cohort studying in the fields of biology, chemistry, and physics at the final stage of schooling, it is of interest to examine the evidence available from 1983-1984 for the SISS in order to determine whether such relationships have been maintained over time. Table 10-1 records, in addition to the proportions of the age group enrolled (expressed in terms of participation rates), the mean levels of achievement of the students in each country for the biology, chemistry, and physics tests. Figure 10-1 provides graphs to show the relationships between mean level of achievement in biology, chemistry, and physics and the proportion of the age cohort involved. These regression lines indicate the expected relationship between participation rate and the mean level of achievement in the science subjects under consideration. These findings confirm those obtained in previous studies. As a

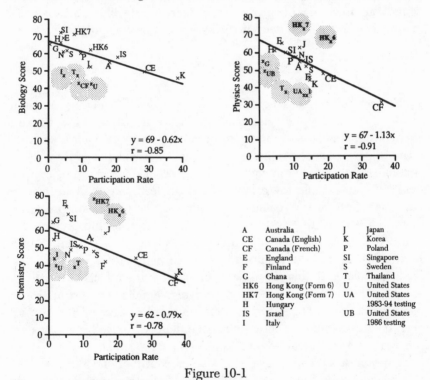

Figure 10-1

Participation Rates and Science Achievement at Terminal Secondary School Level, 1983-1984

consequence of these inverse linear relationships, it must be argued that it is not the mean level of achievement in science recorded for a particular country that is of interest but, rather, whether that country performs above or below the level of expectation indicated by the regression lines drawn in figure 10-1.

In the field of biology, those countries considered to be outliers (that is, performing markedly below expectation) are Italy, Thailand, French Canada, and the USA (figure 10-1). In the fields of chemistry and physics, Hong Kong at both the sixth and seventh form levels clearly is performing well above expectation. However, Italy, Thailand, and the USA perform well below expectation in both the fields of chemistry and physics, and Ghana achieves well below expectation in physics. While these results do not provide evidence of the extent to which countries are preparing an adequate supply of high school graduates in these three fields of science, they do indicate in which particular countries there are significant shortcomings in the achievement of students in their science education programs.

INDEXES OF YIELD

Science achievement yield. The concept of "yield" can be regarded to involve both participation and mean level of performance. Thus, the concept considers how many, in terms of participation, have got how far, in terms of performance. An index of *science achievement yield* for a school system can be obtained by multiplying the science test score by the participation rate. While the calculation of this estimate of yield involves many assumptions that could be questioned, the estimation procedure employed is consistent with the inverse linear relationship presented in figure 10-1. The data on which these calculations are based are given in Keeves (1992) and the results presented in figure 10-2. The yield estimates calculated in this way make allowance for the effects of the different proportions of the age group studying science. The estimates are based on all fields of science, and also the performance of the non-science specialist students, because in some fields very small proportions of the age group are involved, and because in some countries the non-science specialist students make a substantial contribution to the effective achievement yield in science. The high level of achievement yield of the Japanese school system is noteworthy, and the lesser, but still sizeable, yield in Australia is a result of the relatively higher proportion of the age cohort involved. For the USA, while science specialist students make a moderate contribution, the non-science students also make a sizeable contribution. In both cases, the sizes of the contributions are a consequence of the magnitudes of the proportions of the age cohort entailed.

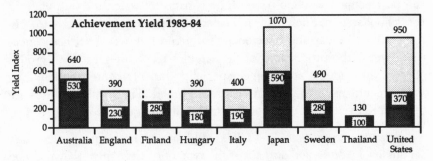

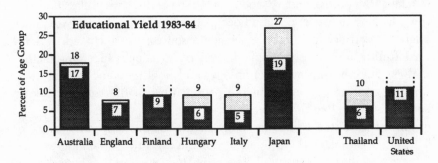

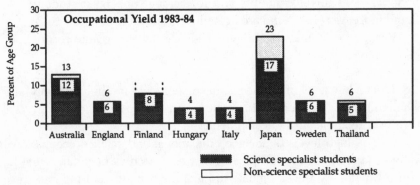

Figure 10-2

Science Achievement and Educational and Occupational Yields, 1983-1984

Science educational yield. Students in the samples tested at the terminal secondary school level were asked whether or not they expected to include any science subjects as part of their further education. It was argued that, if students planned to continue with some study of science at the postsecondary school level, they constituted scientific output from the school system. Such plans indicated a willingness and a desire to continue with the learning of science. To obtain indexes of *science educational yield*, the proportions of the samples responding for both groups (that is, all students and science specialist students) were multiplied by the percentages of the age groups represented. These yield indexes are given in Keeves (1992) and shown graphically in figure 10-2.

Science occupational yield. The students at the terminal secondary school level were asked also about the type of course which they planned to take at the postsecondary school level. The courses were classified into those leading to a science-based occupation, and those leading to a non-science based occupation. The science-based courses included: agriculture, fishing, and forestry; applied science and science including mathematics; engineering and technology; and health sciences, including nursing and medicine. Again the proportions of the samples taking courses in these four areas were multiplied by the percentages of the age groups represented to obtain values of the *science occupational yield*. This index estimates the proportions of the age group planning to train for a science-based occupation. The estimated values of the science occupational yield for all students and for science specialist students are found in Keeves (1992) and are presented graphically in figure 10-2.

From an examination of figure 10-2, the following noteworthy findings emerge:

1. The low level of science achievement yield in Hungary, Italy, and Thailand must be a cause for concern in countries where there is a thrust for technological development.

2. The high levels of science educational and science occupational yield in Japan indicate that Japan was well placed to maintain scientific and technological development. In part, these high yield indexes arose from the plans of some students classified as non-science specialist students to continue with science based courses at the postsecondary school level.

3. Australia had a high level of science educational yield and science occupational yield. Australian students, in general, planned to continue with the further study of science if they were science specialist students.

4. England had a relatively low level of science educational yield and science occupational yield, partly as a result of relatively low retention rates at the terminal secondary school stage and as a result of complete polarization into science and non-science specialist programs at this level.

5. For Hungary, the educational yield and occupational yield were low in 1983-1984. These results were in marked contrast with the high level of aspiration and achievement of the Hungarian students recorded at the 14-year-old level. It would seem that, although the Hungarian students recognized the importance of science for national development, the economic conditions within the country were so depressed in 1983-1984 that the harsh reality of employment and the emphasis on vocational training drew substantial proportions of students away from professional courses in science at the postsecondary stage. This was particularly true for boys.

These findings suggest that a high level of yield is a necessary but not sufficient condition for the advancement of science and technology within a country. Without an ample supply of scientifically and technologically trained personnel, development is not possible. Nevertheless, there must be appropriate economic conditions and organizational structures operating within a country for such development to occur. Japan would appear to have established both the conditions for development and to have trained a substantial supply of scientifically oriented personnel.

Other countries (for example, Australia) appear to have had the necessary supply of personnel, but not the organizational structures and economic conditions required to sustain efforts in this area at a sufficient level. It is clear that the rewards paid to scientists and technologists, as well as the incentives provided through working conditions and support for developmental work, also must be taken into consideration in national planning.

TIME AND ACHIEVEMENT

The data available concerning time spent on learning the separate fields of science are limited because of the difficulties encountered in teasing out the information from the data provided by students and their teachers. Keeves (1992) gives the estimates of time and the mean science achievement test scores, corrected for guessing, in each of the fields of biology, chemistry, and physics. Figure 10-3 presents graphically the relationships between time and achievement.

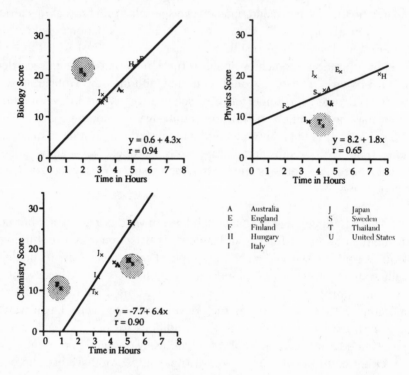

Figure 10-3

**Time and Science Achievement at
Terminal Secondary School Level, 1983-1984**

These results provide evidence that time spent on the study of a subject is a critical factor influencing achievement at the upper secondary school level. In biology, there is a clear relationship between time and achievement, with Sweden an outlier performing well above expectation. The lower level of achievement in biology in Japan, at least in part, is related to the reduced time given to the study of biology compared with chemistry and physics. It should be noted also that a very high proportion of the Japanese sample consisted of girls.

In chemistry, Finland is an outlier performing well above expectation, and Hungary is an outlier performing well below expectation. Hungarian achievement in chemistry warrants further examination.

In physics, Italy and Thailand perform well below expectation. However, the levels of achievement in Italy and Thailand for biology and chemistry were not below expectation. This indicates that, at least in part, the lower than expected levels of achievement in these two countries in biology and chemistry recorded in the examination of relationships between participation rate and achievement were related to less time being spent on the study of these subjects. It has been suggested that the different result in physics might be related to aspects of learning mathematics in these two countries, although there is no direct evidence for this speculative comment.

Only in the field of physics for the USA was information available about both time spent on the study of physics and achievement in physics for an appropriate sample at the Year 12 level. The data recorded for biology and chemistry involved samples defined in alternative ways, and substantial proportions of the students tested were not in their terminal school year. The physics data indicate a level of achievement marginally below expectation after allowance is made for the time spent on the study of physics. In the USA, the teaching of science at the upper secondary level follows a very different arrangement from that occurring in other countries. This leads to lower levels of participation at the terminal secondary school stage in the study of chemistry and in the study of physics in a second year course (as indicated by the USA [B] samples) as well as lower levels of achievement for these groups of students (see table 10.1 and figure 10.1).

ACHIEVEMENT AND OPPORTUNITY TO LEARN

One of the consequences of differences in time allocated to learning science is that, although there is general agreement across countries as to the degree of emphasis to be placed on topics in the fields of biology, chemistry, and physics, it is likely that the students in different countries could have had different opportunities to learn the content associated with the particular items included in the tests. As a result, the relationship between achievement and opportunity to learn should be examined. The science teachers of the students tested were asked to estimate the extent to which their students had the opportunity to learn the science content of the test items. Some ratings were assigned by individual science teachers in a school, whereas other ratings were made by a group of teachers acting in collaboration to produce a common set of ratings for each school in each field of science.

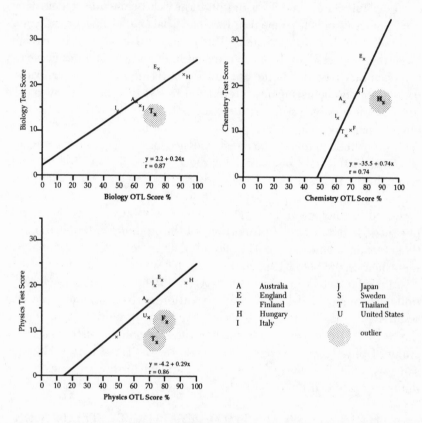

Figure 10-4

Opportunity to Learn Science for Science Specialist Students and Science Achievement, 1983-1984

Two types of ratings of opportunity to learn were used: (1) percentage ratings (in terms of proportions of students); and (2) year level (or grade level) ratings. The percentage ratings were appropriate where it was considered that different students in the school or class group tested had different learning experiences. Where the students in a class group were known to have had identical teaching, then the year level rating was considered more appropriate, as well as being the easier procedure to use. The two sets of ratings were scaled to make them equivalent in terms of a percentage score.

The test scores and the opportunity to learn scores are plotted in figure 10-4. Further information is given in Keeves (1992). In each case, a regression line is drawn. At the science specialist level, outliers include Thailand in biology, Hungary in chemistry, and Finland and Thailand in physics. It would seem likely that the students in the countries identified as outliers had markedly less time in class to master the content that their teachers considered that they had the opportunity to learn. In general, it is clear from figure 10-4 that, where a country has performed at a low level on the science tests, achievement in that country is a reflection of the opportunity to learn afforded.

DOES MORE MEAN WORSE?

An important issue that faces countries which over time increase their retention rates at the terminal grade of schooling is whether standards of achievement have declined as a consequence. In terms of the mean levels of achievement in biology, chemistry, and physics courses, there is clear evidence in the affirmative from the SISS project. The average level of achievement is lower in countries with a higher proportion of the age group involved (see figure 10-1). It is of interest to examine whether the performance of the best students also drops with the participation rate. The best students can be defined as approximately the upper 5 percent of the age group who might be expected to continue with the study of science prior to entry into a science-related profession. However, the proportions participating in the study of biology, chemistry, or physics at the terminal secondary school level in some countries fell below this proportion of 5 percent of the age group, and some lower proportion necessarily had to be used to represent the best students. Postlethwaite and Wiley (1991) give the mean scores for the upper 3 percent of students in biology, the upper 5 percent in chemistry, and the upper 4 percent in physics.

Table 10-2 records the correlations of the achievement of all students in each of the three fields of science with both the percentage of the age group enrolled in academic schools at the terminal secondary

Table 10-2

Correlations between Participation and Achievement of Best Students in Biology, Chemistry, and Physics at the Terminal Secondary Level, 1983-1984

Correlation	All Students' Achievement			Best Students' Achievement		
	Biology	Chemistry	Physics	Biology	Chemistry	Physics
Percent of age group enrolled	-0.59*	-0.47*	-0.34	-0.11	-0.13	-0.12
Percent of age group participating	-0.54*	-0.54*	-0.48*	0.36	-0.10	-0.27

Source: Postlethwaite and Wiley (1991, p. 72)

* $p < 0.05$

school level and the percentage of the age group participating in the study of each field of science. The correlations are all negative and five out of six are statistically significant. Table 10-2 also gives the correlations both for overall enrollment and participation rates in the study of each field with the mean level of achievement of the best students in each field. While five of the six correlations are negative, none is statistically significant. Moreover, the largest correlation between participation rate in biology and mean level of achievement of the top 3 percent of students in biology is moderate and positive (+0.36), but not statistically significant. Because there are no significant relationships between the levels of achievement of the best students in biology, chemistry, and physics and the proportions of the age group retained at school at the Grade 12 level or participating in the study of a subject, it can be argued that the performance of the better students has not suffered significantly as a result of increased retention and participation rates. The nonsignificant but moderate positive correlation of 0.36 for biology students suggests that this subject is drawing on a "pool of ability," presumably of girls in countries where a higher proportion of the age group participate in the study of biology. As a consequence of these results, it is necessary to question the wisdom of holding only small percentages of the age group in specialist science courses at the terminal secondary school level in England, Singapore, and Hungary. Such a question is particularly important in England, where there is known to be a serious shortage of persons trained to work in science-related professional occupations.

SCIENCE ACHIEVEMENT AT LOWER GRADE LEVELS

In examining factors which influence the level of achievement in science of students at the terminal secondary school stage, it is necessary to ask whether the level of achievement at lower grade levels is related to achievement at the final stage. That is, are the effects of learning science cumulative, or is specializing in the study of a particular field of science at the terminal stage sufficient? Three important questions are considered in turn here:

1. Is the average level of achievement in science within a country at the 14-year-old level related to the average level of achievement within that country at the 10-year-old level?
2. What factors influence the growth in average level of achievement in science between the 10-year-old and the 14-year-old levels?
3. Is the average level of achievement in science for a school system at the lower grade levels related to the time given to the study of science and the opportunity to learn the items included in the tests?

Relation between performance at the 10-year-old and 14-year-old levels. Figure 10-5 shows the graph of the average level of achievement

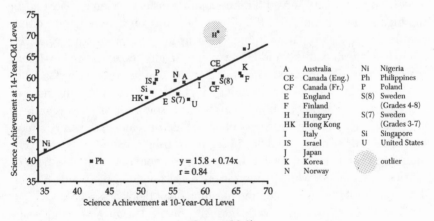

Figure 10-5

Science Achievement at the 14-Year-Old Level and Science Achievement at the 10-Year-Old Level, 1983-1984

of a country's school system at the 14-year-old level plotted against achievement at the 10-year-old level. A general linear relationship is observed. Hungary is an outlier performing well above expectation at 14 years of age given the level of achievement of its students at 10 years of age. Japan also performed at a level above expectation at 14 years of age in spite of its relatively high standing at 10 years of age. The relationship between science achievement at the 10-year-old and 14-year-old levels indicates that the foundations laid in the elementary grades are very important for learning science at the lower and middle secondary levels.

Factors influencing growth in science achievement between 10 and 14 years of age. In a reanalysis of data collected in the first IEA science study, Coleman (1986) drew attention to the fact that some countries taught science at a higher level of intensity at the lower and middle secondary school stages than did other countries. An examination of the growth scores between the 10- and 14-year-old levels for both 1970-1971 and 1983-1984 allows countries to be classified into the two groups as suggested by Coleman. The high-intensity group includes Australia, England, Hungary, Japan, and Thailand, whereas the low-intensity group includes Finland, Italy, The Netherlands, Sweden, and the USA.

A detailed examination of evidence that might account for the differences between the two groups was undertaken in order to identify factors associated with the observed difference in intensity of science teaching. In the high-intensity countries, the three fields of science were taught either simultaneously as separate subjects, or as an integrated subject at the grade levels tested at the middle secondary school stage. In the low-intensity countries at the grade levels tested at the middle secondary stage, not all three fields of science were being taught to all students at the time of testing. In general, alternation occurred each semester or each year. Students in some countries learned only one field of science, as was common in the USA, or at most two fields of science during a semester, as was common in Sweden. In Italy at the Grade 9 level, generally only one field of science was studied; while general science was taught at the Grade 8 level, there was limited emphasis on the teaching of all fields of science.

The effects associated with intensity in the learning of science appear to lie in the fact that, during the period immediately prior to testing, students in the high-intensity countries had an opportunity to learn aspects of science drawn from the three major fields of science—biology, chemistry, and physics—whether science was taught as separate fields or as an integrated subject. However, students in the low intensity countries at the time of testing were studying only one, or at most two, of the three major fields of science. Thus, some students in such countries could have had little opportunity to learn content in one or more of the three fields. In part, this finding reflects the fact that the tests administered contained items drawn from the three major fields of science, together with some items from the field of earth science. The opportunity to learn the science content associated with the items included in the tests inevitably was related to performance.

Time and opportunity to learn in the elementary and lower secondary schools. At both the 10- and 14-year-old levels, the relationship between the time spent learning science as reported by the students and science achievement test scores was examined (see Keeves, 1992, and figure 10-6). The countries differed substantially in the amount of time allocated to the study of science at both levels, and general linear relationships are observed in figure 10-6. In those countries where more time was given to learning science, the level of achievement was higher. At the 10-year-old level, the straight line graph does not pass through the origin. This indicates that there is a substantial level of achievement in science of over fourteen score points associated with no time spent on science learning. This suggests that, in industrialized countries, some scientific knowledge is gained from reading books, magazines, and encyclopedias, and from the mass media including television, as well as from instruction at lower grade levels. There are many other factors in addition to time in science classes that influence science achievement at the elementary school stage, but the importance of curricular time is clear. At the lower secondary school level, the contribution of time spent during earlier years already has been established and the value of the intercept (a=19.5) indicates both the contribution to the learning of science which takes place outside the classroom, as well as the learning of science that occurred in lower grades. At the 14-year-old stage, once again the high level of achievement of the Hungarian students, who performed well above expectation for the time allocated to learning science, provides strong evidence of the benefits of the curriculum reforms in science that took place in the 1970s in that country.

As was discussed in relation to data at the terminal secondary school level, the differences between countries in the time allocated to the study of

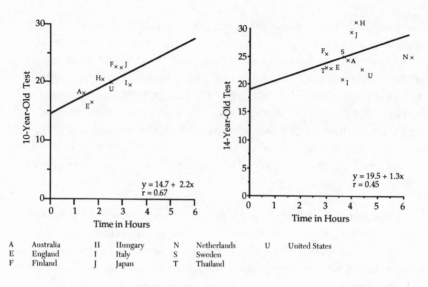

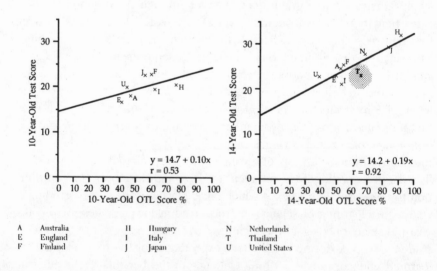

Figure 10-6

Time and Achievement at the 10- and 14-Year-Old Levels, 1983-1984

Figure 10-7

**Opportunity to Learn (OTL) Science and Science Achievement
at the 10- and 14-year-old Levels, 1983-1984**

science at both the upper elementary and lower and middle secondary school levels might be expected to influence the average level of opportunity to learn the items included in the tests. Keeves (1992) reported data on average levels of opportunity to learn and the average levels of science achievement. These results are plotted in figure 10-7 and the existence of linear relationships once again is evident. However, it should be noted that the relationship at the 10-year-old level is not as strong as that at the 14-year-old level when Thailand is treated as an outlier and excluded from consideration.

FACTORS INFLUENCING ACHIEVEMENT WITHIN COUNTRIES

So far, I have examined those factors that account for differences between countries in the outcomes of science education. While the differences in educational policy and practice at the system level influence the average levels of achievement in science of the students within the system, much of the variability that is observed in outcomes occurs at the student, classroom, and school levels. At all age levels, detailed analyses (with the unit of analysis used being between students within schools) indicated that achievement in science is influenced primarily by (a) the status and educational resources of the home, (b) the aptitude and prior learning of the student, (c) the sex of the student, (d) the attitudes and values toward science held by the student, and (e) the motivation of the student (see Keeves, 1992).

At all age levels, detailed analyses also were undertaken between schools, although the results of such analyses in part were confounded by bias resulting from the aggregation of data obtained from students to the school level for analysis. The average level of achievement in science of classroom or school groups was influenced by the following school and teacher factors: (a) the competence of the teacher (but not only in science); (b) the emphasis on inquiry and practical work in the science classroom; (c) the amount of time allocated and the opportunity to learn, where these differ markedly between schools and classrooms within a system; and (d) the average level of attitudes and values of the school and classroom group (see Keeves, 1992).

However, achievement in science is not of itself a valuable outcome if it has been obtained in circumstances in which students do not seek or have the opportunities to continue with the further study of science and train to enter a science-related occupation. The learning of science does not terminate necessarily with the end of compulsory schooling, or the completion of the mandatory study of science during the years of formal education in schools. At a time when participation in programs of recurrent and further education beyond the years of schooling is becoming increasingly necessary and desirable, it is important that students maintain a sufficiently high level of interest in science to plan to continue with the study of science or science-related courses in the years beyond formal schooling. Thus, the factors that influence decisions to undertake training for a science-related occupation and the further study of science beyond the terminal stage of secondary schooling are also of particular interest. However, because of economic circumstances and educational traditions, countries differ markedly in terms of the proportions of the age group held at school to the final years of schooling. These circumstances and traditions can change and are being changed with the gradual growth that has occurred in most countries over the period since World War II toward universal secondary education. Nevertheless, it is important to consider the factors that influence participation in the further study of science at the postsecondary stage within countries. These factors need to be understood and policies need to be developed so that some countries do not suffer the current severe shortages of persons in occupations that sustain scientific and technological development.

FACTORS INFLUENCING FURTHER PARTICIPATION IN SCIENCE

Analyses were undertaken to examine the factors that directly and indirectly influence future participation in science courses after the completion of secondary schooling. The criterion measure in these analyses was a combination of variables formed from students' responses to two questions concerned with: (a) future participation in science courses at the postsecondary level, and (b) expectations of taking a course that provided training for a science-related occupation. Data obtained from students in nine countries were analyzed to determine factors that directly or indirectly influenced further studies in science (see Keeves, 1992, for details of these analyses).

Further study of science was found to be influenced directly by (a) science values (career interest and beneficial aspects of science), (b) aspirations (expected occupation and postsecondary education), (c) amount of science studied (class time and number of science courses taken), and (d) science attitudes (interest in science and ease of learning science).

Several variables in the model used in the analyses had sizeable total effects, although their direct effects were trivial, with the four variables listed above that had sizeable direct effects serving as mediator variables. Thus future study of science was found to be influenced indirectly by (a) home background of the student (parental education and occupation, and number of books in home), (b) aptitude (verbal ability and computational skill), and (c) sex of student.

Only in Hungary did the sex of student (being a male) have significant direct and total effects on further participation. Those boys in Hungary who remained in academic schooling were more likely than girls to continue with the study of science. However, in five countries, sex of student had a significant influence on further participation and operated indirectly through values, attitudes, aspirations, and amount of science studied.

The direct effects of science achievement were relatively slight and were considered significant in only four of the nine countries. However, in Sweden, science achievement had a moderate influence on future participation. The effects of the amount of science studied, which no doubt was dependent on science achievement, was more substantial in eight of the nine countries.

The influence of attitudes and values on decisions to undertake further study of science in all countries except Thailand should be noted. However, in Thailand, most students expressed relatively strong and highly favorable attitudes toward science, so that, with little variation between students on these variables, their contribution to explanation inevitably was statistically nonsignificant. The importance of attitudes and values must be taken into account when plans are developed to attract more students to train for science-related occupations after the completion of their secondary education.

CONCLUSION: SUGGESTIONS FOR IMPROVING SCIENCE EDUCATION

In this chapter the focus is on the outcomes of science education (especially participation, achievement, and attitude) rather than inputs or processes. The chapter is based mainly on data from the Second IEA Science Study (SISS) conducted in twenty-three countries by the International

Association for the Evaluation of Educational Achievement in 1983-1984 among students at the terminal secondary, 14-year-old, and 10-year-old levels. In particular, I have reported, for different countries and different age levels, participation and achievement levels in science, as well as relationships between achievement, time spent in learning science, and the opportunity to learn the content covered by the achievement tests. Finally, I have also reported on factors such as home background, sex, and aptitude which, at the terminal secondary school level, affect both future participation and achievement.

In order to assist in planning improvements in science teaching and to ensure that there is maintained an ample number of high school graduates who wish to participate in the further study of science and to train for a science-related professional occupation, a list of 5 recommendations for policymakers and science educators has been developed from evidence referred to in this chapter. Other suggestions for improving the outcomes of science teaching and learning, but for which the relevant evidence has not been presented in this chapter, can be found in the international reports of the Second IEA Science Study (Keeves, 1992; Postlethwaite and Wiley, 1991; Rosier and Keeves, 1991). The recommendations derived specifically from this chapter are:

1. Because different countries differ markedly in terms of the proportion of students enrolled in science subjects, the study of some science should be mandatory at all levels of schooling.
2. Because science achievement at the 14-year-old level is related to achievement at the 10-year-old level, strong elementary-school science courses should be provided.
3. Because the amount of time spent learning science is a critical factor influencing achievement, adequate curricular time should be made available for instruction in science at all levels.
4. Because of links between science achievement and the emphasis on inquiry and practical work, all science courses should involve an appropriate practical component.
5. Because attitudes influence both participation and achievement in science, positive student attitudes toward science should be fostered.

REFERENCES

Coleman, James S. "International Comparisons of Cognitive Achievement." In *International Educational Research Papers in Honour of Torsten Husén*, edited by T. Neville Postlethwaite, pp. 111-119. Oxford, England: Pergamon Press, 1986.

Comber, L. C., and Keeves, John P. *Science Achievement in Nineteen Countries*. Stockholm, Sweden: Almqvist and Wiksell, and New York: Wiley, 1973.

Husén, Torsten (ed.). *International Study of Achievement in Mathematics: A Comparison of Twelve Countries*, vol. I and II. Stockholm, Sweden: Almqvist and Wiksell, and New York: Wiley, 1967.

Keeves, John P. (ed.). *The IEA Study of Science III: Changes in Science Education and Achievement: 1970 to 1984*. Oxford, England: Pergamon Press, 1992.

Postlethwaite, T. Neville, and Wiley, David E. *The IEA Study of Science II: Science Achievement in Twenty-Three Countries*. Oxford, England: Pergamon Press, 1991.

Rosier, Malcolm J., and Keeves, John P. *The IEA Study of Science I: Science Education and Curricula in Twenty-Three Countries*. Oxford, England: Pergamon Press, 1991.